CONSERVATION OF THE SOIL

CONSERVATION OF THE SOIL

by
A.F. GUSTAFSON, *Ph.D.*
Late Professor Emeritus of Soil Technology,
New York State College of Agriculture,
Cornell University

BIOTECH
2011
BIOTECH BOOKS®
Delhi - 110 035

Gustafson, Axel Ferdinand (b. 1880–)

ISBN 81-7622-212-7
ISBN 978-81-7622-212-9

Published by : **BIOTECH BOOKS®**
1123/74, Deva Ram Park, Tri Nagar,
DELHI – 110 035
Phone: 27383999, (M) 09868812760
E-mail: biotechbooks@yahoo.co.in

Showroom : 4762-63/23, Ansari Road, Darya Ganj,
NEW DELHI - 110 002
Phone: 23245578

Printed at : **Chawla Offset Printers**
New Delhi – 110 052

PRINTED IN INDIA

It is much to be deplored that there is not in our schools a single book to tell the youth what everyone should know concerning the foundations of life in the soil or the conditions under which the generation to which he belongs may pass on the precious heritage to those who are to come after. Such instruction can alone be enforced through the exertions of those who have been brought to see the truth and who are willing to labor for its diffusion.—N. S. SHALER, 1896.

PREFACE

Soil erosion in the United States has been recognized as a menace to agriculture, if not to the population as a whole, for many years. A beginning has been made during the past thirty years and more particularly during the past decade in obtaining accurate information on the losses of soil by erosion and on methods of controlling these losses under varied cropping systems.

"Conservation of the Soil," is presented for the use of landowners and operators, for those concerned with land mortgages, for general readers, for county agricultural agents, and for use in college and high-school courses in soil conservation.

The author's interest in the control of soil erosion dates back to his boyhood. Nearly every spring he assisted with the replanting of corn which had been washed out by heavy thundershowers. For several years he supervised the state soil-erosion control field in southern Illinois and in 1918 published the results jointly with the late Professor J. G. Mosier in *Illinois Experiment Station, Bulletin* 207.

During more recent years he has urged the control of soil erosion in both extension and agricultural experiment station publications. In 1935 he spent a brief leave of absence from his regular duties, as chief soil erosion expert on the Federal Soil Conservation Demonstration Project in southern New York.

In the autumn of 1935 when this book was begun there was no book on this subject. Since then two excellent works have been published: "Soil Erosion Control," by Austin Earle Burges; and "Soil Erosion and Its Control," by Quincy C. Ayres.

Great care has been taken to present complete, correct references and to give credit to the exact sources of data. Pictures that are not the author's own work are acknowledged directly in connection with each illustration.

Hitchcock has been followed for the scientific names of grasses (*Miscellaneous Publication* 200). Other publications of the

U. S. Department of Agriculture have been followed in the main for the names of other plants mentioned in the text.

The author desires to acknowledge the cooperation and helpful suggestions of his colleagues in Cornell University and in the Soil Conservation Service. He will welcome suggestions for the improvement of this book.

A. F. GUSTAFSON.

ITHACA, N. Y.,

CONTENTS

LIST OF ILLUSTRATIONS

CONSERVATION OF THE SOIL

CHAPTER I

INTRODUCTION

"It is now a question whether human culture, which rests upon the use of the soil, can devise and enforce ways of dealing with the earth which will preserve this source of life so that it may support the men of the ages to come. If this cannot be done, we must look forward to the time—remote it may be yet clearly discernible—when our kind, having wasted its great inheritance, will fade from the earth because of the ruin it has accomplished."[1]

These prophetic words written by Shaler more than forty years ago deserve much thoughtful consideration today.

LESSONS FROM THE PAST

History reveals that the Sahara desert once was forested and that the Sahara was inhabited many, many years ago. Asia and South America supply a record of ancient civilizations that lived where desert conditions prevail today. According to Lowdermilk,[2] the disappearance of these civilizations resulted largely from the destructive effects of wind erosion. The Orient teaches the Occident how the people have lived in the same area throughout a period of forty centuries. In parts of the Orient, however, the soil has been lost from large tracts, yet the ruins indicate that the land was once productive and inhabited by thriving populations.

What is the record in this country? According to the United States Soil Conservation Service four million acres have been ruined by water erosion and nine million acres by wind erosion

[1] Shaler, N. S., The Economic Aspects of Soil Erosion, *Nat. Geog. Mag.*, Vol. 7, p. 374, 1896.

[2] Lowdermilk, W. C., Man-made Deserts, *Pacific Affairs*, Vol. 8, No. 4, pp. 409–419, 1935.

within the borders of the United States.[1] In addition, 600 million acres have been badly damaged by wind and water erosion. While these data are subject to revision upon complete investigation they do, nevertheless, suggest the importance of the soil conservation problem to which the American people should give immediate serious consideration in the light of all available information.

The area of land entirely ruined may be regarded as negligible in comparison with the immense area of land badly damaged by erosion. This badly damaged acreage constitutes one-third of the land area of the United States outside large cities and mountainous districts. Much of the more seriously damaged land must soon be abandoned. In other words, under present methods of land utilization the acreage wholly ruined for ordinary agricultural use is increasing at a rapid rate. In the state of New York, alone, land has been abandoned at the rate of approximately 100,000 acres a year during the past several decades. Although the development of transportation with the consequent opening of the eastern markets to the agricultural products of the new, rich, easily tilled soils of the Midwest was a factor, soil erosion was also an important cause of the retirement of these older, hilly eastern lands from agriculture.

Youth does not always seek wisdom from its elders. Nor do youthful nations make the best use of the recorded history of older peoples. As a youthful nation this country needs to consult history in order to learn the results of the methods of soil management employed by ancient, as well as by the older present-day, civilizations and to use the knowledge so gained in conserving its own soil resources.

SOIL FORMATION A SLOW PROCESS

For many years the nation's geologists have been observing the effects of cultivation upon the land. They are no less concerned than are soil scientists about the conservation of the soil for the use of future generations. Professor T. C. Chamberlin,[2]

[1] FULLER, GLENN L., Reconnaissance Erosion Survey Data, *U. S. Dept. Agr., Soil Conservation Service*, SCS-MP-2, 1934 (revised, July, 1935).

[2] CHAMBERLIN, T. C., Soil Wastage, White House Conference of Governors, *Proc.*, Washington, D. C., U. S. Congress 60th, 2d Sess., *House Doc.* 1425, pp. 75–83, 1908.

late head of the Department of Geology of the University of Chicago, at the White House Conference of Governors of the United States in 1908 made the following statement about soil erosion:

"We have as yet no accurate measure of the rate of soil production. We merely know that it is *very slow*. It varies obviously with the kind of rock. Some of our soils are derived from material already reduced to a finely pulverized condition. Such also is the glacial drift, rock flour rasped from the face of the ledge by the glacial file and ground up with old soils. On such a base of half-prepared material, soils may be developed with relative rapidity; but even on these, when the slope is considerable, wind, wash, and cropping remove the surface much too fast for stable fertility. . . . Without any pretensions to a close estimate, I should be unwilling to name a mean rate of soil formation greater than 1 foot in 10,000 years on the basis of observation since the glacial period. I suspect that if we could positively determine the time taken in the formation of the 4 feet of soil next to the rock over our average domain, where such depth obtains, it would be found above rather than below 40,000 years. Under such an estimate, to preserve a good working depth, surface wastage should not exceed some such rate as 1 inch in a thousand years. If one chooses to indulge in a more liberal estimate of the soil-forming rate, it will still appear, under any intelligent estimate, that surface wastage is a serious menace to the retention of our soils under present modes of management. Historical evidence enforces this danger. In the Orient there are large tracts almost absolutely bare of soil, on which stand ruins implying former flourishing populations. Other long-tilled lands bear similar testimony. It must be noted that more than loss of fertility is here menaced. It is the loss of the soil body itself, a loss almost beyond repair. *When our soils are gone, we too must go,*[1] unless we shall find some way to feed on raw rock or its equivalent. The immense tonnage of soil material carried out to sea annually by our rivers . . . is an impressive warning of the danger of negligent practices. Nor is this all; the wash from one acre is often made the waste cover for another acre or for several. Sometimes one's loss is another's gain, but all too frequently one's loss is another's disaster; and the 1,000,000,000 or

[1] Italics are the author's. A. F. G.

more tons of richest soil matter annually carried into the sea by our rivers is the nation's loss."

From the late Professor Chamberlin's observations it is clear that the formation of soil is an extremely slow process as measured in terms of human life: 1 foot of soil in 10,000 years; or 4 feet, the minimum needed for a soil to be a good productive one, in 40,000 years. From these estimates it must be evident to every thinking person that this generation has, and that the many generations to follow will have, the present soil, and none other, upon which to produce the materials needed for food, raiment, and shelter. In other words, mankind is wholly dependent on the conservation and preservation of the soil.

SOIL ORIGINALLY PROTECTED BY FORESTS AND PRAIRIE GRASSES

In 1607 the pioneer settlers arrived in Virginia, and in 1620 the Pilgrims landed at Plymouth Rock. Each group came to a forested wilderness, which was most unpromising as a site for making a home and for producing food. However, as the ax was applied with diligence, a sheltering cabin soon sprang up, and a clearing ample for immediate food and feed production was made (Fig. 1).

Essentially all of the eastern part of the United States was covered with valuable forests of pine, hemlock, and hardwoods and often mixtures of all of these. In the Midwest was a rolling humid prairie area covered with native grasses. The uplands adjacent to the larger streams in this area, however, were occupied by deciduous hardwoods. As the rainfall decreased to the westward, trees practically disappeared, and both north and south of the Ozark Mountains in Missouri the grasses became shorter and thinner with still less and less rainfall until finally they gave up their struggle to clothe the naked earth. There the desert begins. Even the desert produces some vegetation, but it is sparse and affords scant protection for the soil. In the mountains, below the tree line, and on the Pacific Coast the land was occupied by forests wherever the rainfall was sufficient for tree growth.

Before the white man came to North America, nature had established an almost perfect balance or equilibrium between soil, slope, and climate on the one hand and plants and animals on

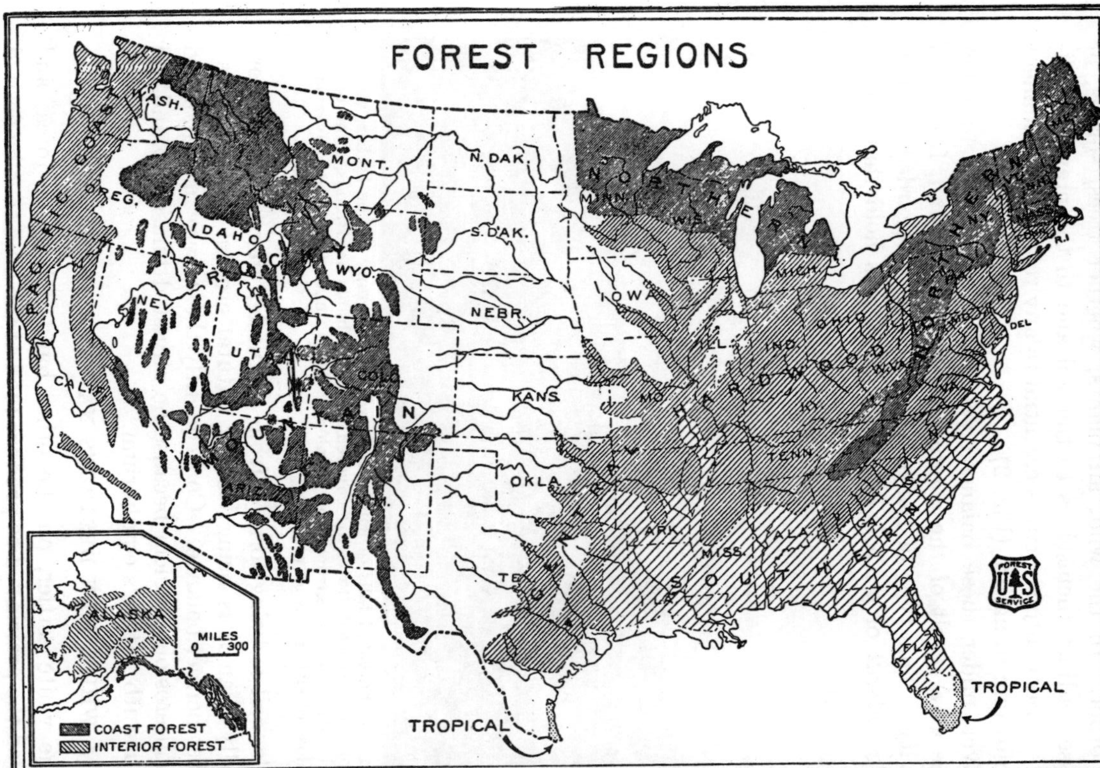

FIG. 1.—Forested areas of the United States. The shaded areas were originally covered with forests. The more hilly and the mountainous areas have remained in forest. (Courtesy of U. S. Forest Service.)

the other. In the wild, all plants, whether trees, shrubs, or grasses, adjust themselves to the soil and to the prevailing climate. Such adjustment is essential to any relatively permanent equilibrium in nature (Fig. 2).

Even under these balanced conditions soil movement takes place as a result of the combined effects of rainfall, frost, and gravity. This type of soil movement may be regarded as inevitable *natural* or *geologic* erosion. Geologists hold that, on the

FIG. 2.—Forested hills of southern New York. Many of the higher, steeper hills of the northeast states are covered with deciduous hardwoods, more or less mixed with pine, fir, spruce, and hemlock. Soils under this forest cover suffer little loss of either soil or water.

whole, soil is being formed from the underlying materials more rapidly than it is being lost by natural erosion under nature's almost perfect balance. Consequently the American Indian had no soil-erosion problem because he adjusted himself to and did not disturb nature's equilibrium.

WHITE MAN DESTROYS NATURAL ADJUSTMENTS

The white settler, on the other hand, felled the forest and plowed the prairie sod to produce food of the kind that he desired for himself and feed for his livestock. The timber was cut from thousands of acres in the northeastern states with no regard for the steepness of the slopes. Much of the wood was burned. The ash or the potash extracted from it was sold in Europe. For the time being, this early potash industry was a most important source of cash income for the settlers. Tanbark was removed from hemlock logs which were left to rot on the land because there was no market for lumber at that time. Since immediate use, rather than conservation of natural resources, was the dominant idea of the time, the now defenseless pioneer must not be censured too severely, since, like present-day farmers, the pioneer *had* to make a living (Fig. 3).

In the South the pine was cut for lumber or "boxed" for turpentine. Boxing weakens the trees to such an extent that many of them are blown down and thus destroyed. Furthermore, the

Fig. 3.—Steep gullied slopes. These steep slopes on the west side of Keuka Lake in central southern New York were cleared many years ago. Gullying, marked by the narrow strips of trees down the slope, soon followed. Many of these gullies are limited in depth only by the presence of the underlying rock.

surface from which the bark and the growing wood is removed in boxing becomes coated with inflammable materials, and many trees are destroyed by fire. During recent years burning has been practiced, all too generally, for the production of early grazing on the vast acreage of sparsely wooded lands in the South.

Frequently the lumbermen in all sections cut the forest clean, burned the brush with the natural undergrowth, and made skid trails and wagon tracks on the slopes, all of which led inevitably to the beginning of erosion and the curtailment of the productive use of these lands (Fig. 4).

After the forest land was cleared much of it was plowed and seeded to crops. Even the steeply sloping land produced good yields of grain and grass for a few years on soils that later proved to be rather unproductive. In explanation, rather than justification, of these agricultural practices it must be admitted that those were the days of oxcart transportation. Food for the city populations was of necessity produced near by. With the construction of the canals and later of the railroads, however, thousands of acres of the steeper lands in the eastern states dropped out of cultivation but not until after the soil had been severely damaged by erosion.

FIG. 4.—Man-made gullies. This well-drained soil in central southern Michigan was originally clothed with hardwoods, largely maple. Plowing, planting, and cultivating parallel to the roads have led to this type of rather severe gullying. Dead furrows caused all of these gullies. In southern Michigan these steep gravelly slopes give good returns when seeded to alfalfa, for which the soil is well adapted.

CROPPING AFFECTS RATE OF ABSORPTION OF WATER BY SOILS

During the earlier years of cropping, the soil of either timbered or prairie areas was loose, since it contained the accumulated leaves and twigs together with the smaller roots and rootlets of the trees or grasses. In this condition it absorbed rainfall readily. In but a relatively few years, however, the natural organic matter decayed, and the more active part of it disappeared from the soil. As a result of these changes, the plowed soil became compact and absorbed water more slowly than it did when first plowed. Consequently much water flowed off over the surface during heavy showers. This *runoff water is that which causes erosion.*

As in the East, so later in the Midwest, much of the forest land was cleared and together with the prairie sod was plowed and planted to corn and other grains. The same procedure was followed to the west and to the southwest. In the drier area still farther westward the native grasses were pastured. As the range became heavily stocked, these grasses were grazed so closely during periods of prolonged drought that they were injured, and in places the vegetation was killed completely.

Furthermore, some years ago *dry farming* was developed in the subhumid section between the humid area and the desert. This type of farming has been widely extended since 1900. Frequent plowing and cropping to wheat destroyed the native grasses. Once this soil protection was gone, the dry, loose, incoherent soil was exposed to rain and wind. Severe soil erosion on the slopes and the dust storms generally in the dry regions inevitably followed.

ESTABLISHMENT OF THE FEDERAL SOIL CONSERVATION SERVICE

Soil erosion is an age-old problem. In this country it dates back to a few years after the native forest was cut off, and the timber and prairie soils were brought under the plow. The problem of controlling erosion became acute in the older sections of the eastern and southeastern states at an early date. Some experimental work was done on the control of soil-erosion losses in that section and later in the Midwest, also. A few years ago the federal government set up a number of well-placed and well-distributed soil-erosion control experiment stations. Valuable data have already been obtained, and as the work proceeds a mass of highly useful information is being developed.

Soils and agronomy extension workers in some states, notably in the South and in the Midwest, have spent no little time with farmers in an effort to establish control of soil erosion. Something of a foundation for future soil-erosion control work was laid during those years.

Hugh H. Bennett of the Department of Agriculture was thoroughly conversant with the seriousness of soil-erosion losses in the country as a result of years of soil work, particularly in the southeastern states. In 1933 he received a grant of money and authority to establish the federal *Soil Erosion Service* in the

Department of the Interior. The work was later transferred to the Department of Agriculture and at that time was broadened into the United States *Soil Conservation Service.*

Under the able leadership of Director Bennett the information possessed by soils men, agronomists, farm-management men, agricultural engineers, and forestry workers has been coordinated and brought to bear on the problems of controlling soil erosion on the farms of the United States.

Under this ably directed organization one or more soil conservation demonstration projects have been established in most of the states. The purpose of these projects is to show farmers and others that soil erosion can be controlled in a practical manner and that the food, forage, and fiber needed for all purposes can be produced at the same time. If the work of New York project number one, which is located in the southern part of the state, is representative of the various projects, a sound beginning has been made in national control of soil erosion in the United States.

Special Reference

Little Waters, by H. S. Person, E. Johnston Coil, and Robert T. Beall, *U. S. Dept., Agr.* 1936.

CHAPTER II

SOIL EROSION

A distinction is sometimes made between natural erosion which is the result of geologic forces and that which is the result of man's efforts to obtain a living from the land. Natural or *geologic* erosion is so slow in terms of human life as to be wholly unnoted by all except the geologist. The fact that soil as we think of it[1] covers the surface of the earth today shows that, over the ages past, soil has been formed from the underlying material faster than it has been removed. Obviously, if this were not true the land surface of the earth today would consist of bare rock and not of soil. Geologic erosion is not detrimental to man's well-being and is wholly beyond his control. In fact, geologic erosion is unimportant in comparison with man-induced erosion which is sometimes called *accelerated erosion*. For these reasons the simple term *erosion* will be used on these pages to designate the erosion that has been induced by man in his management of the land. Man can, however, exercise a large measure of control over many phases or stages of man-induced soil erosion. This is the menace with which all of us are deeply concerned.

Soil erosion results from the action of two distinct factors or forces, water and wind, and by the two combined as waves on shore lines. Together these agencies effect erosion over a very large proportion of the land surface of the earth.

EROSION BY WATER

Comparatively level, open sandy or gravelly soils and peats absorb a normal rain as it falls. Sloping soils of the heavier types, such as silt and clay loams, take up water more slowly than do the lighter soils such as sandy and gravelly loams. If rain falls on relatively level land faster than the soil can absorb it,

[1] More or less decomposed and disintegrated rock material mixed with more or less organic matter and, under favorable climatic conditions, capable of sustaining plant life.

water remains on the surface as ponds. On slopes, on the other hand, water that is not taken up quickly by the soil is lost as runoff over the surface. It is this runoff water that carries away soil or that causes erosion.

Sheet Erosion.—As water passes over a smooth slope, it flows along in a sheet of more or less uniform depth. On areas of unprotected sloping soil the sheet of water picks up soil particles and carries them along down the hill. More particles will natu-

FIG. 5.—Surface soil lost by sheet erosion. Although it does not appear steep the slope along the gully is 7 per cent. This fine field was abandoned because of the loss of surface soil. Note alluvial fan in foreground. Photograph taken near Anna, Va.

rally be washed from soft bare soils than from those protected by vegetation. As long as the sheet of water is of essentially the same depth, and if the condition of the land is relatively uniform, the removal of soil is approximately the same from all parts of an area having a similar degree of slope. This moderately uniform removal of surface soil is called *surface* or *sheet erosion* (Figs. 5 and 6).

Sheet erosion has damaged millions of acres of sloping lands throughout the United States. In places on the gentler slopes only an inch or two of soil has been lost, but on steeper slopes or on more easily eroded soils all of the surface, or plow, soil to a depth of 6 or 7 inches has been lost. In still other places varying quantities of the subsurface, in addition to the surface, soil have been washed away.

Wherever sheet erosion has removed more than half of the original surface 7 inches of soil and has cut into the subsurface in places the natural productivity of the soil has been materially

FIG. 6.—Sheet erosion in southern New York. This erosion was the result of a heavy shower, unofficially reported as 2 inches in 30 minutes. The slope from the right is 10 per cent. Water came down the rows, deposition occurred where the slope flattened out.

reduced. Shallow-surfaced soils suffer greater reduction in productivity from such losses than do the deeper-surfaced ones. And shallow soils underlain by a tight or impervious stratum or by bedrock are far more seriously injured than are those soils which have an open, well-drained subsoil.

Another phase of sheet erosion is that which accompanies the thawing of frozen soil in winter or in early spring. As thawing proceeds on normally wet, fine-grained soils, the soil and water form a thin mud which runs down the slope. If, as often happens, much rainfall accompanies thawing, the mobile mud is thinned so that it flows away rapidly, much the same as does ordinary runoff water.

A somewhat similar situation obtains where a heavy clay stratum or a tight or impervious layer occurs within 12 to 18 inches of the surface. In the spring the surface soil readily becomes saturated because of the slow percolation of water through the impervious layer. Raindrops beating on a saturated surface easily loosen the fine and medium-sized soil particles which consequently are readily washed away. Long-continued rains sometimes produce a similar condition on well-drained, medium-grained soils. Here, too, beating rains bring soil particles into suspension and carry them off along with soluble plant nutrients and organic matter.

Sheet erosion causes great economic loss which cannot be computed in dollars and cents, because the surface soil is more productive and hence more valuable than the subsurface or the subsoil material. In addition to this loss, active organic matter and readily available plant nutrients are carried away by sheet erosion. On sloping lands which have received commercial fertilizer and farm or green manures, sheet erosion removes part of the added plant nutrients and organic matter.

Rill Washing or Shoestring Gullies.—Sheet erosion leads inevitably to other forms of erosion unless areas of sheet-washed soils are given complete protection, such as may be afforded by a favorable change in methods of cropping. Tiny gullies a few inches in depth and width frequently develop on unprotected slopes of recently cultivated soils. These gullies are referred to as *rill-washing* or *shoestring* gullies and, if neglected, grow into large gullies. As a rule, however, these tiny gullies are erased by subsequent tillage during plowing, seedbed preparation, or cultivation. Thus the surface is smoothed again, and when thus managed, rill washing becomes in effect sheet erosion. Rill washing may be regarded as a transition stage between sheet erosion and gullying (Fig. 7).

Gully Erosion.—Gully erosion differs from sheet erosion in that gullying is the removal of soil from a relatively narrow belt or strip down a slope. Any depression, such as an open furrow, depressions between the rows of crops, a wheel track, a cow path, a mole run, a woodchuck or prairie-dog run, or, in fact, any depression, however slight, down a slope may lead to the beginning of a gully. Like fire, a gully grows by what it feeds upon, and, like fire out of control, gullies are extremely destructive. The damage by gullying does not take place so rapidly as does that from fire; yet, when a gully is actually cut 12 feet deep across a field during a single night, the power of water to work rapidly on a steep slope must be recognized (Fig. 8).

The kind and condition of the soil have a distinct bearing on gully formation. Loose, open, well-drained sloping soils gully

FIG. 7.—Rill washing on summer fallow land in Washington. This type of erosion takes place on land summer fallowed for wheat. Upon cross tillage these little gullies are filled and so become in effect sheet erosion. The thin sward of wheat for summer pasture in the foreground "stopped all visible erosion damage to the land." (*Courtesy of U. S. Soil Erosion Service and Washington State Agr. Exp. Sta.*)

rather easily when water is concentrated on them. A heavy or tight subsoil intensifies sheet erosion under many conditions, but such a layer often checks the rate at which gullying takes place. A moist, compact clay stratum is more or less slippery and rubbery, and the water with its burden of soil, sand, gravel, and stones readily slides over it with little abrasive effect. Rounded or water-worn pebbles possess little ability to erode such a clayey subsoil. Flat sharp-cornered stones, on the contrary, scratch, cut, and gouge off some of the clay in passing over it and thus in time gnaw it away. Some clays check and crack into large cubical granules, which on being loosened by the water fall from the sides into the gully and are carried away by the current. Under this soil condition, gullies may grow in width very rapidly.

A compact, impervious subsoil may act in somewhat the same manner as clay, but such a layer seldom enjoys the protective effect of the wet clay.

FIG. 8.—Gullying in south central New York. These gullies were formed during a two-day rain July 7 and 8, 1935. The largest gully is as much as 12 feet deep in places. Water which was concentrated at the top of the ridge by beans and potatoes came through the woods and wrought havoc on the slope below. This farm was in the heavy rainfall area (see Fig. 36).

Along the Mississippi, Missouri, and some of the smaller rivers in the United States is a wind-laid soil material, called *loess*, which possesses some rather definite characteristics. A similar formation is found in China and in other parts of the world as well. In this country the loess is from a few feet to 100 or even 200 feet in depth. Loess consists of an unusually uniform-sized soil material which contains a very low percentage of clay or colloid material. The reason for this condition is that the wind which deposited the loess naturally carried the finest material as dust far beyond the present location of the deep loess. Deep loess possesses a characteristic perpendicular cleavage, and a section of it in a gully or on a road or stream bank stands up like a rock cliff. When a stream undercuts deep loess it caves in or falls over in slabs extending from the bottom to the top of the section. Owing to undercutting, deep loess and loesslike materials erode rather badly once the protecting surface soil has been cut through (Fig. 9).

Fig. 9.—Characteristic erosion in deep loess. The soil absorbs water but it does not resist erosion once the surface soil has been cut through. Owing to the vertical cleavage of this soil, it falls off in slabs when it is undercut by water. Photographed, 1910.

Loess or loesslike material underlain by water-bearing fine sand is subject to severe gully formation. The water rolls the sand out from under the overlying loess. The pressure of the loess on the sand may aid the water in starting the movement of the sand and in keeping a steady flow of it going out from under the loess. As the sand moves away, the overlying loessial material caves in and rolls down into the gully and is itself in turn carried away. A gully of this type grows not only as a result of the runoff water's coming in at its head but by removal of the underlying sand and the resulting caving in of the overlying material. Control of this type of erosion is unusually difficult (Fig. 10).

Waterfall Erosion.—Water causing sheet erosion in flowing down over a slope may cut a hole on an unprotected spot. Such a hole often becomes a small waterfall. As the water goes over the face of the fall it tends to undercut at its base, and the top soil falls in. Thus in time the waterfall eats its way up the slope and forms a gully of considerable size unless the waterfall is brought under control in the early stages (Fig. 11).

Landslide or Slip Erosion.—Streams especially at flood stage often undercut their banks, particularly on curves. Thus the

Fig. 10.—Gully erosion in central Illinois. This soil is underlain with sand which is being carried out continually by spring water. As the sand goes out from under the surface soil it caves in. This form of erosion is difficult to control. The planting of willow cuttings and water-tolerant grasses in the bottom of the gully, and other grasses, vines, shrubs, and trees on the sides should afford a large measure of control. The drainage area is not more than two or three acres. Photographed, 1910.

Fig. 11.—Waterfall erosion. Water concentrated in the depression which was a head-land furrow 25 years ago is rapidly cutting its way back up the hill in this excellent pasture soil in western Illinois. This gully is on the farm on which the writer was born. It was photographed during a visit in April, 1936.

slope of the clifflike stream bank may be left steeper than the angle of rest when wet, although in dry weather the bank is perfectly stable. In the early spring or following a long period of heavy rainfall such banks become saturated, and in that condition landslides occur. The soil and the rock material slip into the stream and dam it temporarily but only to be cut away by the current in a short time. Highway cuts sometimes are left too steep for stability in a wet condition and suffer frequent

FIG. 12.—Slip erosion in southern New York. This soil had been stable since it was formed. Unusual conditions attending the heavy rains of July 7 and 8, 1935 brought about this situation. Approximately 40,000 tons of soil material slid down the hill on the forenoon of the eighth following two days of heavy rains, leaving a hole 400 by 100 feet and covering 5 acres of beans on productive soil below. Slope of hill, 40 to 42 per cent.

landslides which cause traffic delays and entail heavy recurring expense for the removal of the landslide material.

While landslides may not be erosion, strictly speaking, they are closely associated with the action of water and sometimes frost in addition. Water has a buoyant action on the subsoil material (Fig. 12). Moreover, if the subsoil contains much clay, this, when wet, acts as a lubricator and facilitates the action of gravity on the mass. Steep sides of cuts, whether made by streams or made by man for highways, railroads, or canals, present a real problem. This problem is particularly difficult if the soil is clayey or if it contains strata bearing water at any season of the year. In a sense, small landslides play a large part in gully formation.

EROSION BY WIND

Wind erosion, as already noted, has caused serious soil losses in many sections of the United States. Medium-coarse-textured soils, such as medium, fine, and very fine sands and sandy loams; medium-textured soils, such as loams and silt loams; and fine-

FIG. 13*a*.—Dust storm in Oklahoma on April 14, 1935. (*Courtesy of H. A. Daniel.*)

textured soils, such as clayey silt loams, silty clay loams, clay loams, and clays, are all subject to movement by the wind under favorable conditions. It is only when dry and loose, however, that these soils are easily blown about by air currents (Figs. 13*a* and 13*b*).

Sand Dunes.—Dry, fine sand is easily moved by wind of moderate velocity, but a light shower holds the sand particles together and thus checks their movement. A very brief period of bright sunshine, however, quickly evaporates the water that holds together the sand grains on the immediate surface, and a light breeze starts the sand to drifting again. In the spring, sand grains from the surface of sandy loams may be loosened and separated more or less from the finer binding material. Such conditions lead to local dust storms and to movement by salta-

FIG. 13*b*.—Dust storm in South Dakota in April, 1934. (*Courtesy of Mrs. M. C. Jorgenson, Watertown, S. D.*

FIG. 14.—Sand dunes on the Atlantic coast of Long Island, New York. Migrating sand can be anchored by means of vegetation.

tion, a leaping or jumping movement, of sand particles by the wind (Fig. 14).

Silt Dunes.—Loams and silt loams blow most when they are very dry, especially after being more or less pulverized by the processes of seedbed preparation and cultivation. Silt dunes which formed years ago are a common phenomenon in deep-loess areas (Fig. 15).

Clay Dunes.—When granulation of the finer soils, especially if well supplied with organic matter, occurs, as it frequently does over winter, blowing of the granules takes place in dry periods in winter and spring. The clayey soils when granulated drift as do the heavier silt loams. These soils, however, tend to break into lumps or clods during cultivation and in that condition are less subject to blowing than are the coarser soils. Black clay

FIG. 15.—Silt dunes in central Illinois. The silt was supplied as outwash at the close of the glacial period.

loams in the Midwest granulate to a marked degree upon freezing and thawing over winter. This action is representative of what happens to heavy soils in general if they are well stocked with organic matter. Fall-plowed fields have no vegetative protection throughout the winter; consequently, if such land is but scantily covered with snow, much blowing of the clay-loam granules occurs.

Coffey describes clay dunes[1] in Texas formed by the blowing of clay granules in precisely the same manner as that in which sand dunes are formed.

Blowing of Peat Land.—Cultivated peat, or muck, lands in the northern part of the United States are subject to blowing when

[1] COFFEY, GEORGE N., Clay Dunes, *Jour. Geol.*, Vol. 17, p. 754, 1909.

the immediate surface of the soil is dry. In the case of peat formed largely from wood, the small pieces of preserved wood are picked up and carried along by saltation. The granules of peats formed from reeds or sphagnum are moved in the same manner. Much material thus blown from cultivated fields is deposited in ditches, on highways, or in other places where it is not wanted.

Man Encouraging Erosion.—Thus all but universally in this country has the white man seriously disturbed, if not destroyed, nature's equilibrium. The resulting unbalance has exposed the soil over millions of acres to wind and rain. And both of them have made the most of an opportunity long denied them by nature herself.

EROSION BY WAVES

Ocean and lake shore lines are being cut back by wave erosion which results from the action of water with wind as the driving force. The rate at which the land is cut away varies with the material of which the shore line is composed and in a measure with the direction of the prevailing winds. Obviously a coast line composed of hard rock yields to wave action more slowly than does one composed of soft shale or of unconsolidated materials. And waves that beat upon the shore at right angles to the direction of impact cut more rapidly and drag the material cut loose farther out from land than do waves that strike the shore at an acute angle. The tides intensify wave action on ocean coast lines more in some situations than in others. All shore lines, however, in time yield to the eternal gnawing of the waves.

Special Reference

Deserts on the March, by Paul B. Sears, *University of Oklahoma Press*, Norman, 1935.

CHAPTER III

RESULTS OF SOIL EROSION

Some results of soil erosion have been indicated briefly in the foregoing pages, but additional data may be of service.

RESULTS OF EROSION BY WATER

Based on their extensive investigations, Humphreys and Abbot estimated that the Mississippi River[1] carried in suspension and deposited in the Gulf of Mexico each year soil material equivalent to a column 241 feet high over a base of 1 square mile. If the material rolled and pushed along on the bottom of the river was included and was based on estimates with projected levees completed, the height of the column became 315 feet. This is equivalent to 7 inches of soil over approximately 350,000 acres. These estimates were published in 1861. The quantity of material transported by the Mississippi River is undoubtedly much greater today, because many acres of forest have been cleared and together with a large acreage of prairie soils have been brought under the plow. In addition, the natural protective organic matter in both forest and prairie soils has been destroyed since that time. And all of these changes have led to a marked intensification of erosion.

More recently Bennett[2] has estimated the sediment carried into the Gulf of Mexico by the Mississippi to be 429 million tons a year or the equivalent of the surface soil over 429,000 acres.

Dole and Stabler[3] estimated that 783 million tons of soil material, one-third of it in solution, is washed from the lands of the United States and carried into the sea each year. The

[1] HUMPHREYS, A. A., and H. L. ABBOT, Report on the Mississippi River, *Corps of Engineers U. S. Army, Prof. Papers* 13, pp. 146–147, 1861, reprinted 1876.

[2] BENNETT, H. H., Soil Erosion a National Menace, *U. S. Dept. Agr., Circ.* 33, Part I, p. 17, 1928.

[3] DOLE, R. B., and H. STABLER, Denudation, *U. S. Geol. Survey, Water-Supply Paper* 234, p. 83, 1909.

solid sediment in this immense mass of material is equivalent to the weight of the surface soil over 513,000 acres of land.

Chamberlin, as previously quoted, estimated 1,000 million tons or more as the annual loss of soil carried into the sea from the lands of the United States. This mass of material is equivalent to the surface 7 inches of soil over one million acres of land.

Most Valuable Soil Material Carried Away.—A high proportion of these immense quantities of material consists of the richer, more productive surface soil and its fresh or active organic matter. Some of it does come from gullies cut through the top or plow soil, but by far the major part of it is surface material.

The soil material carried into the ocean consists of colloidal matter, clay, silt, and the finest grades of sand. This is the portion of the soil, particularly the colloidal matter and clay, from which crops can extract the plant nutrients required for their growth more easily than from the coarser soil particles. Consequently, the value to crops of the soil material lost to the sea is far greater than is apparent from a mere consideration of its actual weight or total quantity.

Much surface soil and subsoil material as well is washed from fields and deposited on the flatter part of slopes or at the base of slopes, on bottom lands or flood plains of streams, in swamps, and in lakes and reservoirs. Obviously the cultivated land loses many times as much soil every year as is actually carried into the sea. Bennett has suggested that the loss from fields and pastures may be a hundred times as much as that which actually enters the sea.[1] No entirely satisfactory basis exists for an accurate estimate, yet soils men do know that the total soil loss by erosion from sloping cultivated fields is very great. And this loss certainly is many, many, times as great as the total loss to the sea.

Immense quantities of plant nutrients, also, are carried in solution to the ocean every year. In so far as this represents the soluble material in underground waters reappearing as springs or directly in streams, the loss cannot well be avoided, regrettable though it may be. Large quantities of plant nutrients are removed annually from cultivated lands by the runoff water. This loss, although it cannot be separated from that removed from the soil by true underground water, is largely avoidable.

[1] Bennett, H. H., Soil Erosion a National Menace, *U. S. Dept. Agr., Circ.* 33, Part I, p. 4, 1928.

The loss is serious, since the runoff water carries away the more readily available plant nutrients, some of which may have been applied as purchased fertilizers. Crop yields are materially reduced by these losses of nutrients. At least for some decades to come, the nutrients lost can be replaced with phosphorus from the large stores of phosphorus in the various domestic and foreign deposits, with potash from Searles Lake in California, from the immense underground deposits in New Mexico and foreign countries, and from domestic industrial wastes and with nitrogen from the Chilean deposits or by-product nitrogen and from the inexhaustible supply of atmospheric nitrogen. This atmospheric nitrogen may be fixed electrically or by legumes for the use of nonlegume crops.

While the loss of nutrients is regrettable and is costly to the farmer, it is nothing compared with the loss of the soil material itself. Nutrients may be replaced, but how can the best of the soil be put back on the fields? In the main, the cream of the soil lost goes all the way to the sea, and some of the intermediate-sized material is deposited on flood plains of rivers, creeks, and smaller streams. None of this valuable soil material can ever be returned to the fields whence it came. It is lost to the farms and largely to mankind, forever.

Damage by Erosion to the Soils of the United States.—In Tables 1 and 2 are shown the acreage and the percentage of total area of the United States exclusive of large cities and water areas that have been affected by erosion (Fig. 16). These figures show the perfectly staggering damage that the nation's soil resources have suffered up to the present time, and unfortunately the process is continuing, practically unabated. Sheet erosion, as previously indicated, is the usual form of the most widespread damage. Slightly more than one-third, 34.8 per cent, of the "total area" of the United States has lost from approximately 2 to 5 inches of its original surface soil by sheet erosion. And more than 5 inches has been lost in the same way by an additional tenth, 10.1 per cent, of the total area exclusive of large cities and water. One sixth, 17.7 per cent, has been gullied severely in addition to sheet erosion, and a relatively small acreage (4,162,092), 0.2 per cent, has been ruined by gullying and sheet erosion acting together. It is interesting to note that 30.3 per cent of the total area has undergone little or no erosion. Under this classification is

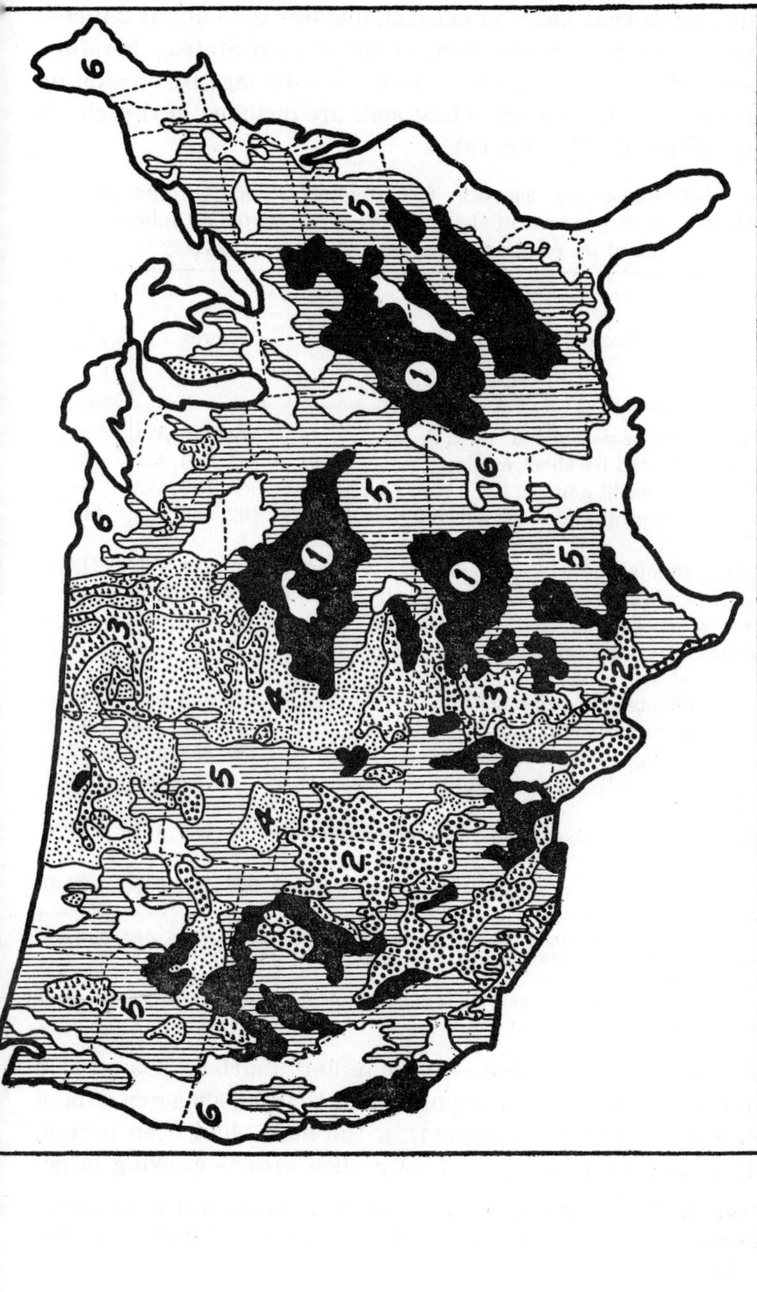

FIG. 16.—General erosion conditions in the United States. 1. Severe sheet and gully erosion. 2. Moderate to severe erosion, includes mountains, mesas, canyons, and badlands. 3. Moderate to severe wind erosion, some gullying locally. 4. Slight wind erosion, moderate sheet and gully erosion. 5. Moderate sheet and gully erosion, serious locally. 6. Erosion unimportant, except locally. (*Courtesy of U. S. Soil Conservation Service, August,* 1936.) Redrawn by H. O. Buckman.

included lands that have, in general, lost less than about 2 inches of topsoil. Less than one-third of the United States, therefore, has been relatively free from considerable damage by erosion up to this time. The country's best soils are included in this classification (Figs. 17, 18, and 19).

TABLE 1.—SUMMARY OF EROSION CONDITIONS IN THE UNITED STATES*
(More than 25 per cent of the land has been affected as indicated)
(Based on reconnaissance erosion survey, 1934)

Erosion conditions	Acres	Percentage of total area†
Area with little or no erosion	576,236,371	30.3
Total area affected by sheet erosion	855,260,347	44.9
¼ to ¾ of topsoil gone	663,199,473	34.8
Over ¾ of topsoil and some subsoil gone	192,060,874	10.1
Total area affected by wind erosion	322,181,740	16.9
Moderate wind erosion (with some gullying)	233,321,336	12.2
Severe wind erosion (with some gullying)	79,659,052	4.2
Destroyed by wind erosion (with some gullying)	9,201,352	0.5
Total area affected by gullying‡	864,818,281	45.4
Occasional gullies (with sheet erosion)	523,351,168	27.5
Severe gullying (with sheet erosion)	337,305,021	17.7
Destroyed by gullying (with sheet erosion)	4,162,092	0.2
Rough mountain lands, canyons, and rock outcrop	144,768,315	7.6
Total area exclusive of large cities and water	1,903,176,620	100.0

* FULLER, GLENN L., Reconnaissance Erosion Survey Data, *U. S. Dept. Agr., Soil Conservation Service*, SCS-MP-2, 1934 (revised, July, 1935).

† Based on last item in table.

‡ In the western states much gullying is normal geological dissection, on which erosion may be, but is not necessarily, active.

Soil Losses by Erosion.—In mapping Fairfield County in South Carolina the soil surveyors found that 90,000 acres of land which once had been cultivated, in the main, had been permanently ruined by gullying.[1] This gullied area is growing larger

[1] CARR, M. E., F. S. WELSH, G. A. CRABB, R. T. ALLEN, and W. C. BYRES, Soil Survey of Fairfield County, South Carolina, *U. S. Dept. Agr., Bur. Soils*, 1914.

TABLE 2.—EROSION CONDITIONS IN THE UNITED STATES BY CLASSES*
(More than 25 per cent of the land has been affected as indicated)

Erosion class	Acres	Percentage of total area†
Little or no erosion, less than ¼ of topsoil lost	576,236,371	30.3
With occasional gullies	94,354,965	5.0
With frequent gullies	16,532,663	0.9
Moderate sheet erosion, ¼ to ¾ of topsoil lost	97,860,716	5.1
With occasional gullies	296,406,164	15.6
With frequent gullies	174,460,705	9.2
Severe sheet erosion, more than ¾ of topsoil lost; may have lost subsoil also	6,760,021	0.4
With occasional gullies	60,633,625	3.2
With frequent gullies	108,819,243	5.7
Destroyed by gullying	4,162,092	0.2
Moderate wind erosion, some topsoil gone; local accumulations	124,378,797	6.5
Moderate sheet and moderate wind erosion	18,780,563	1.0
Moderate wind erosion with occasional gullies	16,189,063	0.9
With frequent gullies	1,271,274	0.1
Moderate sheet, moderate wind, occasional gullies	42,879,814	2.3
With frequent gullies	16,838,947	0.9
Severe sheet, moderate wind erosion, occasional gullies	3,153,190	0.2
With frequent gullies	9,029,688	0.5
Severe wind erosion, much topsoil gone, destructive accumulations	54,776,159	2.9
With moderate sheet erosion	5,902,905	0.3
With severe sheet erosion	192,390	‡
With occasional gullies	5,776,630	0.3
With frequent gullies	278,592	‡
With moderate sheet erosion, occasional gullies	2,840,644	0.2
With frequent gullies	7,229,015	0.4
Severe sheet erosion, severe wind erosion, occasional gullies	248,698	‡
With frequent gullies	2,424,019	0.2
Extreme wind erosion, losses and drifting too severe for cultivation	8,702,102	0.5
With occasional gullies	78,375	‡
With frequent gullies	420,875	‡
Mesas, canyons, badlands, rough mountain land	10,049,727	0.5
Barren mountains, areas above timber line	130,135,388	6.9
Scablands, shallow soils with frequent rock outcrop	4,583,200	0.3
Total area, exclusive of large cities and water	1,903,176,620	100.0

* Reconnaissance Erosion Survey Data, *U. S. Dept. Agr., Soil Conservation Service*, SCS-MP-2, 1934 (revised July, 1935).

† Based on last item in table.

‡ Less than one-tenth of 1 per cent.

FIG. 17.—Gullying following sheet erosion in southern Ohio. This slope varies from 18 to 22 per cent. Canada and Kentucky bluegrass along with wild white clover are struggling with the elements for control. A little encouragement would assure success to the vegetation.

FIG. 18.—Gullying in southern Illinois. According to the owner, who is shown in one of the gullies, this pasture was in cultivated crops only a few years before the photograph was taken in the spring of 1911.

year by year, and the problem of its control is growing more difficult with the passing of the seasons.

Duley[1] believes that most of the worn-out lands of the country are worn out because their surface soil has been washed away rather than because they have been depleted of plant nutrients by cropping. Even the more gently rolling lands in Missouri, he said, have lost so much surface soil that they are much less

Fig. 19.—A gully in southeastern Iowa. This gully formation follows the loss of the surface soil which is underlain with incoherent sand. Control methods may be applied to such gullies with success.

productive than when they were first brought under the plow a century ago.

"Clay points" are a common occurrence in western Illinois. The brown soil has been eroded away, leaving the yellow clayey subsoil. This subsoil is now being cropped in place of the original surface soil which was much more productive than are the clay points. On these areas corn and oats seldom produce more than half the yield which is normally obtained from areas which still possess their original surface soil.

[1] Duley, F. L., Controlling Surface Erosion of Farm Lands, *Missouri Agr. Exp. Sta., Bull.* 211, 1924.

In Doniphan County, in northeastern Kansas,[1] the soil surveyors found that as an average for the county 6 inches of surface soil has been lost by erosion from the good productive lands. A virgin timbered area had from 12 to 24 inches of loose rich surface material over its clay subsoil. At the same time adjacent cultivated land had lost all of this loose rich soil and in some places part, and in others all, of the subsoil as well. In places, 4 feet of soil had been washed off, exposing the underlying limestone.

Rockie and McGrew[2] report the effects of a heavy rain on July 30, 1931, over an area of about 50 square miles, near Colfax, Wash. This heavy rain was not unusual and was no heavier than occurs in many sections of this country.

On bare fallowed slopes the water accumulated as a sheet 18 inches deep and in places even deeper. The water rushing down the slope carried soil and litter which was piled against the fence at the foot of the slope to a depth of 18 inches or more.

Detailed measurements were made on a small tract which represented erosion conditions over the area. From 1 to 3 inches of soil was lost from 20.3 per cent of this measured area, from 3 to 6 inches from 35.2 per cent of it, and from 6 to 10 inches from 44.5 per cent of this representative area. The authors are convinced that this summer-fallow land lost an average of 2 inches of soil as a result of this one downpour. This loss is equivalent to 275 tons of soil an acre. On that basis 20,000 acres of fallow land in this severe-storm area in the Palouse hills lost 5.5 million tons of its best surface soil during one rainstorm.

Effect of Erosion on Land Values.—In south-central New York nearly 7 inches of rain fell in a little more than 48 hours in July, 1935. Soil-erosion conditions on five farms having a total of 810 acres were mapped shortly before the rain. The same land was remapped soon after that downpour. The results of erosion during that one rain are shown in Table 3.

These data indicate what can happen to rolling farm lands during very brief periods. The rain increased the acreage of severe sheet erosion by more than one-third, more than doubled the area of land with frequent gullies, and multiplied by four the

[1] Bennett, H. H., *U. S. Dept. Agr. Circ.* 33, Part I, p. 9, 1928.

[2] Rockie, W. A., and P. C. McGrew, Erosive Effects of Heavy Summer Rains in Southeastern Washington, *Washington Agr. Exp. Sta., Bull.* 271, 1932.

acreage on which the gullies are too deep to cross with farm implements. In addition to this damage 17 acres, or one-fiftieth of the area of these five farms, was covered by deposits of coarse, almost valueless material.

TABLE 3.—THE EFFECT OF A SINGLE RAIN ON SOIL-EROSION CONDITIONS AND LAND VALUES IN SOUTHERN NEW YORK*

Soil-erosion conditions	Before, acres	After, acres
Little or no erosion	17.8	11.3
Slight sheet erosion	336.9	257.8
Moderate sheet erosion	302.6	313.3
Severe sheet erosion	151.9	208.7
Very sevére sheet erosion	0.0	3.8
Occasional gullies	98.6	208.1
Frequent gullies	36.9	83.1
Gullies too deep to cross with farm implements	5.0	22.0
Covered by deposition	0.0	16.9

* HOWE, F. B., and H. R. ADAMS, Soil Erosion in New York, *Cornell Agr. Ext. Bull.* 347, p. 14, 1936.

A glance at Fig. 8 shows what happened in one day to another good hill farm. These data represent severe erosion conditions. From them it is clear that long-term loans on relatively erosive rolling farm lands need to be protected by the putting into effect of a thoroughgoing soil-erosion control program as a condition for the granting of loans. Farm values are not necessarily permanent on more or less steeply sloping lands.

Destructive Deposition of Eroded Materials.—Up to this point consideration has been given to the loss resulting from the removal of soil from farm lands. Consideration follows of instances of losses resulting from the deposition on good land of coarse material brought down by streams.

In Steuben County, near the New York-Pennsylvania state line, the heavy rains of July, 1935, caused heavy losses to valley farmers. One small stream, which drains 4 square miles of land, brought down silt, sand, gravel, and stones from the ridge lands to the bottom lands. The larger stones were dumped on good productive land as an alluvial cone or fan, at the mouth of the V-shaped gorge occupied by the stream. It is estimated that 15,000 tons of stones and finer materials was spread out over an area of more than 5 acres of good valley soil. The largest stones were deposited nearest to the upland, and the depth of stone was greatest there. Both size and depth of stone feathered out as the debris was carried farther out into the valley. Next to the stones came the gravel followed by the sand, and farther out from the hills the finer materials were deposited. The land covered with small stones and gravel was estimated to be about 5 acres in addition to that covered by large stones. Fully an

FIG. 20.—Stones and gravel deposited by a small stream. The stones brought down by a small stream covered several acres of good crop land as a result of one long rain in southern New York in July, 1935.

equal area was covered with fine soil material. While this latter land will eventually become productive, the 5 acres covered with stones and coarse gravel have been rendered permanently unproductive except as the stream may later cover these coarse materials with finer deposits. This represents material already mentioned as being removed from the farm lands but never reaching the sea. Almost innumerable examples of this nature may be found generally in hilly sections throughout the country (Figs. 20 and 21).

In 1934 in the same county a somewhat similar difficulty was observed. In this instance potato rows sloped directly toward a country road. The water from a heavy shower collected and ran down the slope between the rows into the road-drainage ditch. At a sharp bend in the road the water was diverted on to a wooded slope. With proper spreading of the water such diversion should be a safe procedure. In this instance, however, the drainage failed to spread and tore down the steep slope, making large gashes wherever the trees were so far apart as not to afford sufficient protection. The result was a pile of rocks, estimated at

Fig. 21.—Sand and silt at outer edge of deposit shown in Fig. 20. This field of fine timothy was covered in part by sand and silt to a depth of more than a foot. Note the bean cultivator at the left.

about 100 loads, dumped in a public road at the base of the slope. The finer materials had been carried over the road into a swamp where they did no damage but are of no present value.

In Vermont and other New England states the flood of November, 1927, covered many acres of productive bottom lands with stones, gravel, and sand in the same manner. This coarse material is generally unproductive.

Bennett[1] states that streams in Mississippi which once were navigated have become so clogged with sediment from near-by cultivated fields that steamboats can no longer navigate them.

[1] Bennett, H. H., Soil Erosion a National Menace, *U. S. Dept. Agr., Circ.* 33, Part I, p. 8, 1928.

So many sand bars have been built up in the streams that navigation has been impossible for some years.

Bennett[1] reports further that during a rainy period in the southwestern part of the United States the soil washed from a burned-over hillside did tremendous damage in an orange grove. Where the deposit was more than 12 inches deep it had to be hauled off, and where less than a foot in depth the debris was raked back from the trees. This work is estimated to have cost the owner $1,200 an acre.

Silting of Reservoirs.—In the older section of the country, water power was of extreme importance in the early days for the operation of local flour and grist mills and sawmills. Previous to the clearing of forests from the hills the flow of streams was relatively steady throughout the year as compared to their present flow. While the stream flow was uniform, the water-power supply, also, was relatively uniform at all times of the year.

It is notable that so many old water-power plants continue in operation in the area around the Adirondack Mountains. In the immediate vicinity of Ithaca, N. Y., on the other hand, eleven dams have been built at various times. Of these eleven reservoirs, seven have been filled with silt and consequently have no storage capacity at present. Two of the dams have been washed away. Four of the eleven dams are on the same creek, and three of them are still functioning to a degree. They are serving more as diversion dams, however, than as reservoirs. The only other of these eleven reservoirs still in service is the Ithaca city-water reservoir. This dam was built in 1910 with an original capacity of 357 million gallons of water. By 1935 its capacity had been reduced to 276 million gallons, or by 23 per cent in twenty-five years. In 1925 a silt basin was constructed above the reservoir at a cost of $13,642. Since that time 150,000 cubic yards of silt has been removed from the basin at a cost of $1,400. It is estimated that removal of the silt in the main reservoir in order to restore it to its original capacity would cost $100,000, or the total cost of restoration would be at the rate of $5,000 a year[2] since the dam was built. Calculations show that in the

[1] BENNETT, H. H., Soil Erosion and Flood Control, Lecture 1 before Graduate School, U. S. Dept. Agr., Jan. 30, p. 9, 1928.

[2] Information supplied to the writer by Dr. John Lamb, Superintendent of the U. S. Soil Conservation Experiment Station near Ithaca, N. Y.

twenty-five years of its existence this reservoir has received a total of 60 tons of soil from each acre of cultivated land in its drainage area.

The Schoolfield dam near Danville, Va., was completed in 1904.[1] The dam, which is 1,150 feet long and 35 feet high, originally formed a reservoir of 540 acres of an average depth of 17 feet. In sixteen years 432 acres of the reservoir had been covered with silt and coarser materials to an average depth of 19 feet. A considerable portion of it had been built up above

FIG. 22.—Schoolfield dam on the Dan River, Virginia. Note the island of silt out in the center of the reservoir area. Photographed June 1, 1935. (*Courtesy of U. S. Soil Conservation Service.*)

the original surface of the water. All that now remains of the reservoir is the narrow stream channel. During the past 16 years the gates in the dam have been opened regularly in an effort to get rid of some of the deposited silt. The dam now serves only as a diversion dam and does not produce a reservoir of much consequence (Fig. 22).

In the southern Piedmont, reservoirs are silting very rapidly. Thirteen reservoirs as reported by Eakin[2] have been completely filled in an average period of less than 30 years. Rapid silting is in progress in Texas, Oklahoma, Arizona, Georgia, and North and South Carolina, as may be seen in Table 4.

[1] KEIL, P. F., Two Centuries of Accruing Tragedy along the Dan River, *U. S. Dept. Agr., Soil Conservation*, Vol. 1, No. 7, p. 4, 1936.

[2] EAKIN, H. M. *The Land, Today and Tomorrow*, Vol. 1, No. 2, p. 4, 1934.

From the data in Table 4 wide variations are noted in the rate of silting. The Lake Michie and Lloyd Shoals reservoirs will not be completely filled, if it is assumed that present conditions continue, in less than about three hundred years. Corresponding assumptions and figures are about forty-one years for the life of the Waco and forty-eight years for the Spartanburg reservoir. All of the other reservoirs studied in that area are intermediate between these extremes. Based on the figures in column 8 of Table 4 the average life of all of these reservoirs will be approximately one hundred years. It should be pointed out, however, that these figures for the estimated life of the reservoirs based on the data in Table 4 represent the entire period required to fill

TABLE 4.—RESERVOIRS SURVEYED BY SOIL CONSERVATION SERVICE, 1935*

Reservoir	Location	Period	Age, years	Original capacity, acre-feet	Storage per square mile of drainage area, acre-feet	Annual silt accumulation per 100 square miles of drainage area, acre-feet	Annual depletion of storage, per cent	Total depletion of storage to date of survey, per cent
Lake Michie	Durham, N. C.	April, 1926	8.75	13,457	79.16	27.69	0.35	3.07
University Lake	Chapel Hill, N. C.	June, 1932	2.92	2,076	76.92	87.9	1.13	3.29
Greensboro	Greensboro, N. C.	February, 1923	11.5	2,870	39.80	31.3	0.78	9.05
High Point	High Point, N. C.	August, 1927	7	4,220	72	60.8	0.84	5.87
Lake Concord	Concord, N. C.	March, 1925	10.16	1,179	310.49	200.8	0.65	6.57
Spartanburg	Spartanburg, S. C.	May, 1926	8.17	2,700	29.3	62	2.1	17.15
Lloyd Shoals	Jackson, Ga.	December, 1910	24.33	107,702	78.24	26.5	0.34	14.58
Rogers	Rogers, Tex.	September, 1922	12	164	300	568	1.9	23
Lake Waco	Waco, Tex.	April, 1930	5	42,223	25.41	61.78	2.43	12.16
White Rock	Dallas, Tex.	1910	25	19,540	171.41	140.83	0.82	20.54
Guthrie	Guthrie, Okla.	October, 1920	14.5	3,232	243.03	234.81	0.97	14.01
Boomer Lake	Stillwater, Okla.	March, 1925	10.25	2,794	306.11	182.01	0.59	6.05
Elephant Butte	Hot Spring, N. M.	January, 1915	20.25	2,638,860	100.29	76.97	0.76	15.5
San Carlos	Coolidge Dam, Ariz.	October, 1928	6.33	1,247,999	92.15	43.05	0.46	2.93

* BENNETT, H. H., Report of the chief of the Soil Conservation Service, p. 41, 1935.

them completely. The efficiency of the reservoir, of course, will be greatly lowered after its storage capacity is reduced by one-half. And in case of a city reservoir it must be abandoned and replaced by another at great expense, or it must be cleaned, also at great expense, long before it is filled completely with silt.

According to Bennett,[1] the Harding reservoir in Orange County, California, was completely filled with soil debris by a 24-hour rain in November, 1926. Burning of the bush protection on the steep hillsides caused that disastrous flood. It is reported that on adjacent areas where the bush was undisturbed this rain caused no trouble whatsoever; in fact, the streams ran clear instead of being a mud torrent such as came off the burned-over land.

An average annual deposit of 1 foot of silt in thirty stock-water reservoirs in the Southwest is reported by Talbot.[2] Such deposits would reduce the life of these reservoirs to less than fifteen years. Such a threatened loss and interference with the livestock industry of the region certainly demands action leading to holding the silt in place. Instances of ruinous deposition might be multiplied indefinitely from observations and from the literature, but these suggest the problem.

Any reservoir, pond, or lake, natural or man made, becomes filled in time because of the irresistible downward movement of soil material in response to gravity. In nature undisturbed it is estimated that filling will occur at a rate of about 1 foot in three hundred years. In comparison, man-made reservoirs are silting with highly destructive rapidity. Such rates of filling raise a serious question as to the economy of reservoirs.

Desirable Deposition of Eroded Materials.—In the spring of 1905 while mapping soils in the Mississippi bottom in Pike County, Illinois, across the river from Hannibal, Mo., the writer observed an apparently rather unusual situation. A square 40-acre field of coarse yellow silt loam was bounded on three sides by heavy clay soil. Soil boundaries seldom follow such arbitrary man-made lines as the fences around a 40-acre field. Upon inquiry, it developed that the owner of the field five years previously had cut the levee of the small creek which flowed down

[1] BENNET, H. H., Soil Erosion a National Menace, *U. S. Dept. Agr., Circ.* 33, Part I, p. 17, 1928.

[2] TALBOT, B. W., Range Watering Places in the Southwest, *U. S. Dept. Agr., Bull.* 1358, p. 44, 1926.

one side of his field and maintained a low dike on the lower sides of the field. This permitted the creek to spread its burden of yellow silt over his field. In those five years the clay soil had been buried to a depth of from 4 to 6 feet by yellow silt, or an average of about 2,000 tons of soil an acre was deposited over this field each year. The field had been transformed from a heavy, intractable, unproductive clay to a moderately productive, easily tilled silt loam; and, furthermore, because of the change in the stream this land was less subject to overflow than it had been before it was built up.[1]

Another stream in the same county spread its load of the same yellow silt from the deep-loess upland over a field of timothy nearly ready to harvest. The grass checked the current and thus brought about immediate deposition of the silt. By actual measurement the silt was found to be 6 inches in depth as the result of a single heavy but not unusual rain for western Illinois.

In Saint Clair County, Illinois, near East St. Louis, 700 acres of a lake had been filled to a depth of from 10 to 30 feet since the adjacent deep-loess uplands had been cleared. By 1906 a considerable area of this newly formed or "made" land had been built up to a point high enough to be sufficiently well drained to be cultivated. And in that year 120 acres of it was planted to crops.

A similar instance is reported by Bennett.[2] In the Missouri River bottom in Doniphan County, Kansas, a farmer built a dike around a 40-acre field in order to intercept the material that small streams were bringing from the nearby uplands. Within ten years the soil had been built up to a level with the top of the dikes, a depth of 5 to 7 feet. Approximately 1,200 tons of silt a year was deposited within the dikes, but all the while the finer material was being lost to the Missouri River. While this newly made soil was less productive than that which it covered, it seems probable that its deposition in this position was wholly favorable in the long run. The land will now be overflowed less frequently than formerly, and, moreover, a smaller total acreage of the more productive soil was covered in this way than would

[1] GUSTAFSON, A. F., Dumping Soil into the Sea, *Farm and Fireside*, June 5, pp. 5–7, 1915.

[2] BENNETT, H. H., Soil Erosion a National Menace, *U. S. Dept Agr., Circ.* 33, Part I, p. 10, 1928.

have been buried if the streams had been allowed to follow their own natural courses.

In May, 1906, while engaged on the Illinois soil survey the writer made notes on instances of erosion and methods of prevention. The following instance is drawn from these unpublished notes. The Baltimore and Ohio Railroad had a trestle (No. 329.07 in May, 1906) west of Caseyville in St. Clair County, Illinois. Persons who had lived in the locality for many years stated that large loads of hay were driven under this trestle fifteen years previous, or about 1891. To quote the notes: "It [the driveway] is now filled to within 10 inches of the stringers. They [the residents] state that the driveway has filled up from 10 to 12 feet during the past five years." That is filling at the rate of about 24 inches a year.

Near by, on the right of way, Western Union Telegraph poles 50 feet long were set 5 feet in the soil and 45 feet above, according to the railroad section boss who had worked there for twenty-five years. In May, 1906, for quite a long distance the poles stood with only from 10 to 12 feet aboveground. The wash, therefore, was from 30 to 35 feet deep. The poles were set twenty-five years before (1906), indicating an average deposit of from 14 to 16 inches annually during that quarter century.

RESULTS OF WIND EROSION

According to data in Table 1, 9 million acres of land in the United States have been ruined by wind erosion, and 80 million acres have been badly damaged in the same manner. In some states wind erosion has not been and is not now a serious problem. Iowa, for instance, has less than 4,000 acres that have been ruined by wind erosion. Illinois has 30,000 acres that have been ruined or severely injured by wind erosion.

In central Illinois, dune sand lies in the midst of highly productive land. Wind movement of the sand is a problem even though the exact location of the sand within the dune area may be a matter of minor concern, because the sand itself is mainly unproductive wherever it may be. Dune sand holds a continual threat against good land. The wind may break through such protecting vegetative cover as may be present and start the dunes to moving. Once moving, the sand does not stop on sand but spreads out over productive lands. Even when a dune after a period of stabilization, superimposed on good soil, again moves on, some sand remains behind. This lagging sand consists of material that is too coarse to be moved by ordinary strong winds. Nothing short of a gale can move it. The result is that this coarse sand forms a desert pavement, in effect. It is unproductive and consequently is worthless. This coarse sand forever holds imprisoned beneath it the rich soil which once produced excellent yields of numerous crops. On one central Illinois farm with which the writer is wholly familiar some 15 acres of produc-

FIG. 23.—Blowing sand covering productive soil. The spring winds are carrying dune sand over onto and covering good productive soil with sterile sand. In central Illinois, spring of 1936.

tive brown fine sandy loam has been covered by coarse unproductive sand during the past twenty-two years (Fig. 23).

In the Adirondack Mountain foothills are thousands of acres of sands and very sandy loams much of which unfortunately was cleared. In places the sand has blown off leaving a boulder pavement which protects the underlying material from further movement. To establish vegetation on drifting sand is not easy of accomplishment (Fig. 155).

Blowing Sand Injures Crops.—The blowing of soil material frequently does severe damage to crops. In central Illinois, for example, cowpeas are drilled on sandy loams and loamy sands. At 1 to 3 inches in height the plants are exceptionally tender and susceptible to injury. Angular sand particles hurled swiftly against them cut the outer layers of the tender stems. The plants die as a result of one or two days of much severe abrasion by the blowing sand. In some seasons a second planting is killed, and a third is then required in order to produce a crop. Thus the wind causes heavy loss to the farmer. Other crops, also, are injured by the blowing sand, but legumes appear to be peculiarly susceptible to this type of injury.

Damage from Blowing in Dry Areas.—In the drier areas of the Southwest, West, and Northwest various crops are injured by the abrasive action of sand and other soil particles or granules. Sometimes the roots of plants are exposed by the blowing away of the soil, and as a result the crop is ruined or its yield greatly reduced. In times and places of severe soil blowing, fences are buried, ditches filled, and roads covered or obliterated. Soil is drifted about farm buildings to such an extent as to require weeks of labor to remove it in order to make the yards and buildings usable again.

Crops of wheat and hay plants have been severely injured by being covered with in-drifted soil. Well-established alfalfa with an abundance of top left over winter protects the soil against blowing. The alfalfa cannot, however, prevent soil from drifting onto it from adjacent unprotected lands. In the long-term view one man's loss may be another's gain, but the farmer whose productive field of alfalfa has been smothered out by in-drifted soil from the fields of shiftless or unfortunate neighbors may not always take the long-term view. Moreover, if the land with the dead alfalfa on it is not reseeded to a crop that will protect the soil against blowing, that soil may be blown on to still other fields and thereby extend the loss to other farmers.

The soils of the West may be less "raw" underneath than are the subsoils of other sections of this country, but few areas exist that can stand many successive removals of the top 6 inches of soil; and few farmers want 6 inches of soil deposited on their fields from adjacent lands. Among other disadvantages such deposits are likely to cover field roads and clog established surface drainage systems. It is unquestionably to the highest advantage of the greatest number of farmers, and in the highest interest of the nation in the long run, that all soils be held securely in their original place. For this purpose their original location may be defined as that which was occupied before man disturbed the native protecting vegetative cover.

Wind Damage on Peat Lands.—Tender vegetable crops on well-drained peat or muck lands are subject to injury in the same way as are cowpeas on the sandy lands of central Illinois. Peat formed from wood has innumerable bits of more or less preserved, angular, hard pieces of wood. In other places hard sharp granules of peat cut the same as do bits of wood. It is

FIG. 24.—Wind injury to crop on peat in New York. The blowing peat had killed a crop of onions on the bare area in the foreground. Lettuce has been planted in order to obtain some income during the season. The picket fences and the willow windbreak in the background afford considerable protection.

these granules that are blown against plants and that injure them (Fig. 24).

Owing to its formation in lake or swamp positions peat usually possesses inadequate, if indeed any, natural drainage. Consequently, the maintenance of drainage, usually with open ditches, is absolutely imperative. The surface peat is blown into and clogs the lateral ditches. The main ditches usually have a flow of water sufficient to carry away the in-blown peat. In time the peat shoveled out builds up an undesirable levee along the lateral ditches. Loss from the blowing of peat results in direct injury to the crop and from the use of labor in cleaning the ditches at a time when crops require the attention of all of the regular farm hands.

Blowing of Heavy Soils in the Cornbelt.—Ordinarily, the cornbelt states are regarded as not having a soil-blowing problem. Heavy silt loams and clay loams, however, undergo granulation upon freezing and thawing, particularly when plowed in the fall. These granules are readily rolled or hurtled along by saltation in strong winds. The soil lodges in field or highway drainage ditches which become clogged to varying degrees. The extent of damage depends on the length of time that granulated soils are exposed owing to lack of snow cover, on the proportion of the winter months that the soil is dry and frozen, on the velocity of the wind, and on the duration of high-velocity winds. Blowing of this type of soils occurs in dry periods during early spring as well as in the winter months.

RESULTS OF WAVE EROSION

Water causes erosion under the force of gravity, wind causes erosion directly, and waves cause erosion wherever land and

FIG. 25.—Wave erosion on the Atlantic coast of Long Island, New York.

water meet. All ocean and lake shores and in a measure the shore lines of the larger rivers, in addition, are a battleground between land and water. Great variation exists in the rate of wearing of shore lines. Rock coasts wear away very slowly in comparison with shore lines composed of soil material (Fig. 25).

Shaler[1] relates that the waves have worn back the shore line of Nantucket Island as much as 6 inches some years and that the coastline of Martha's Vineyard is being cut back at an average rate of 3 feet a year.

Two feet a century is regarded as normal wearing of shore lines by wave action. The cutting back of shore lines by 200 feet since the latest ice age is regarded as rapid wave action from the geological point of view. It is reported that the soft rock of the southern and eastern shores of England have worn down so greatly as to reduce materially the area of many parishes. It is estimated that the land lost to the sea during the past nineteen hundred years is not less than 100 square miles, or 64,000 acres.

Special Reference

Silting of Reservoirs, by Henry M. Eakin, *U. S. Dept. Agr., Tech. Bull.* 524, 1936.

[1] Shaler, N. S., The Economic Aspects of Soil Erosion, *Nat. Geog. Mag.*, Vol. 7, pp. 328–338, 1896.

CHAPTER IV

INFLUENCE OF PRECIPITATION AND LATITUDE ON SOIL EROSION

Among the more important natural factors affecting soil erosion are precipitation, latitude, slope of the land, and the kind of condition of the soil.

INFLUENCE OF PRECIPITATION

Total Precipitation for the Year.—Soil erosion is influenced to a marked degree by the total precipitation for the year. If a large proportion of the year's precipitation falls as slow, gentle showers, the water will be entirely absorbed by many soils. Under these conditions there is no runoff and no erosion. If much of the annual precipitation falls as snow on unfrozen soil, and if the blanket of snow becomes deep enough, little freezing of the soil occurs. Under these favorable conditions the water resulting from the thawing of the snow soaks directly into the soil without runoff, and, consequently, little or no erosion takes place.

If, on the other hand, the soil is frozen deeply under a thick cover of snow, and, more particularly, if thawing of the snow is accelerated by heavy warm rains, most of the resulting water must run off over the surface. Essentially these conditions prevailed in some areas in the northeastern states in March, 1936. And the unprecedentedly high flood waters at Pittsburgh and Johnstown, Pa., at Hartford, Conn., and at all the other cities and villages and on farms along the larger rivers in the open country were the result of that combination of circumstances. Ways of alleviating floods resulting from such conditions are dealt with in Chap. XVII, The Control of Floods (pages 295 to 303).

Distribution of Precipitation throughout the Year.—Uniform distribution of the rainfall throughout the year, that is, approximately the same quantity each month or, better still, each week, would be favorable for control of erosion. Unfortunately, however, precipitation does not come uniformly in many sections of the United States. Two different unfavorable conditions obtain: one is the moderately heavy rain which continues over a number of consecutive days; and the other, the extremely heavy rain of short duration.

After several days of moderate to heavy rainfall the soil becomes saturated. Since the soil can take up more water but very slowly, the continuing rain must run off over the surface. Rains have continued over a period of 14 days, with, of course, large variations in intensity. Such prolonged rains inevitably cause runoff and lead to erosion on sloping cultivated land.

In many sections much of the rain comes as the extremely heavy shower, sometimes called a cloudburst; in fact, the characteristic type of growing-season rainfall in some areas is the sudden, heavy thunder-shower. In this type of rainfall the water comes down so rapidly that even those soils which are loose, open, and in good tilth are unable to take it up as fast as it falls. Runoff takes place with consequent erosion on cultivated slopes and on other lands inadequately protected by vegetation.

Rainfall Intensities.—A close relation exists between erosion and the excess of rainfall over the absorptive capacity of the soil. However, if the intensity of the rainfall is high; that is, if rain falls five or ten times as rapidly as the soil can take it up, heavy runoff and severe erosion result on unprotected slopes regardless of the possible absorptive capacity of the soil.

Under average farming conditions most soils escape serious erosion from average rainfall. Man cannot influence the time or control the rate of precipitation. He can, however, keep the soil in such condition and select and so place his crops as to hold to a minimum the erosive effect of the heaviest showers or of the high-intensity rainfall. Loss by erosion cannot be entirely avoided in the case of those heaviest downpours, such as occur but once in fifty or one hundred years. Surely it is wise management to be prepared at all times for heavy rains such as are likely to come on the average once in twenty-five years.

Flood control, as well as soil-erosion control problems on the farm, ought to be attacked on the basis of the precipitation intensities. Pittsburgh, Hartford, Johnstown, and other cities will surely plan to protect themselves to a degree of reasonable safety from fifty-year floods.

Long-time detailed rainfall records are of equally great value in planning for soil-erosion control on arable lands or for flood control for the protection of the cities in the valleys. From these continuous records the rainfall intensity frequency has been worked out. For example, eastern North Dakota, Illinois, western Texas, West Virginia, and western Massachusetts may expect 1.85 inches of rain in 30 minutes once in twenty-five years, as may be seen in Fig. 28. For this area, then, the rainfall intensity frequency is 1.85 inches in 30 minutes once in twenty-five years. The probability is strong that on the average that quantity of water will fall in 30 minutes once in twenty-five years in those areas in the future, because according to the records such has occurred in the past.

The following charts by Yarnell (Figs. 26 to 35) show the rainfall intensity frequencies for the United States based on the long-time precipitation records of the U. S. Weather Bureau.[1]

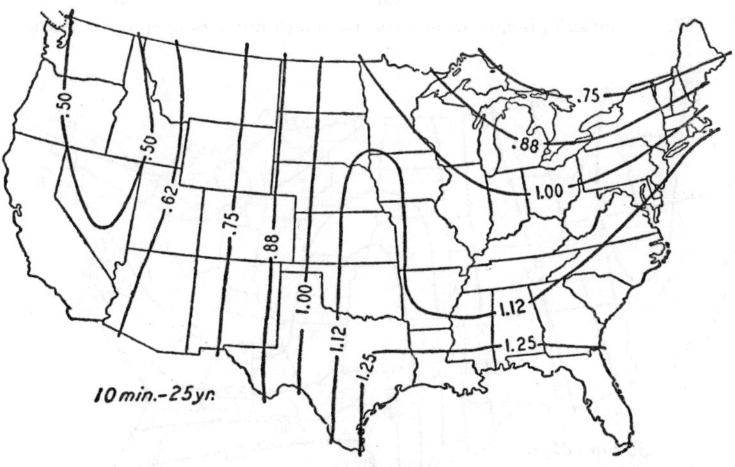

FIG. 26.—Rainfall in inches to be expected in 10 minutes once in 25 years.

From these charts one may determine the quantity of rain to be provided for over short intervals and for 24-hour periods. It appears conservative to plan agricultural operations and such public works as highway sluices, terrace outlets, and man-made stream channels on the basis of the highest intensity, which,

[1] YARNELL, DAVID L., Rainfall Intensity-frequency Data, *U. S. Dept. Agr., Misc. Pub.* 204, 1935.

according to the records, is to be expected once in a quarter of a century. It may be more economical, however, to build highway sluices on the basis of fifty- rather than twenty-five-year intensi-

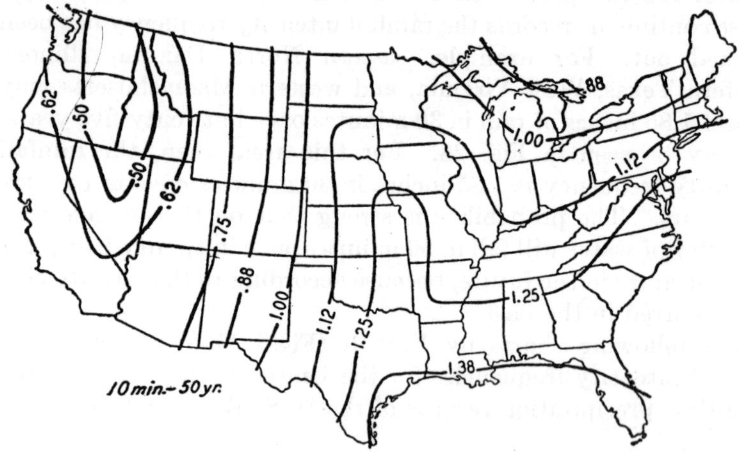

FIG. 27.—Rainfall in inches to be expected in 10 minutes once in 50 years.

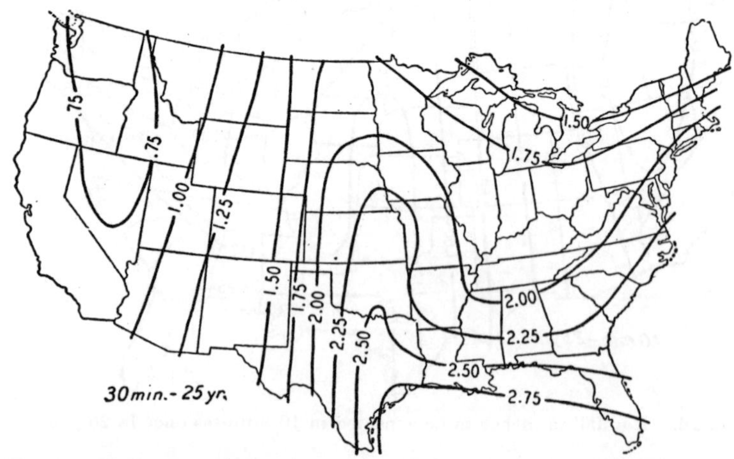

FIG. 28.—Rainfall in inches to be expected in 30 minutes once in 25 years.

ties of rainfall. As already suggested, it is probable that cities would save money by providing for their own protection against floods based on the fifty-year intensities. The fifty- and one hundred-year intensity frequencies together with those for periods shorter than twenty-five years are given by Yarnell. This information is invaluable to soil conservationists and to soil-conservation and flood-control engineers.

South-central New York experienced the heaviest rainfall ever recorded there on July 7 and 8 in 1935. In the area from

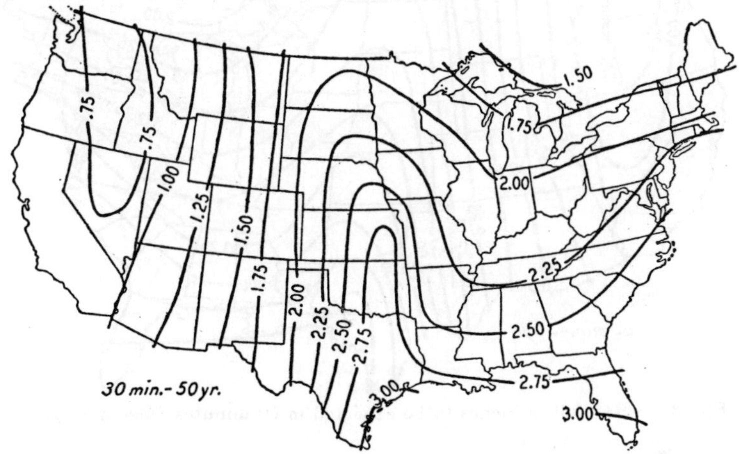

Fig. 29.—Rainfall in inches to be expected in 30 minutes once in 50 years.

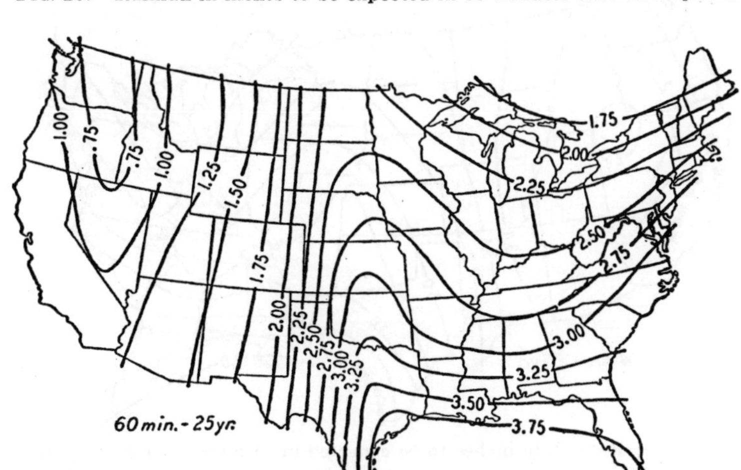

Fig. 30.—Rainfall in inches to be expected in 60 minutes once in 25 years.

northern Steuben to northern Delaware County new 24-hour precipitation records were established at that time. The extremely heavy rainfall extended over an area of more than 100 miles east and west and from 15 to 20 miles north and south. From 9 to more than 10 inches of water fell during the two-day period July 7 and 8, 1935. In many places much of it fell during 24 hours. In some places the damage was not serious, while in others the

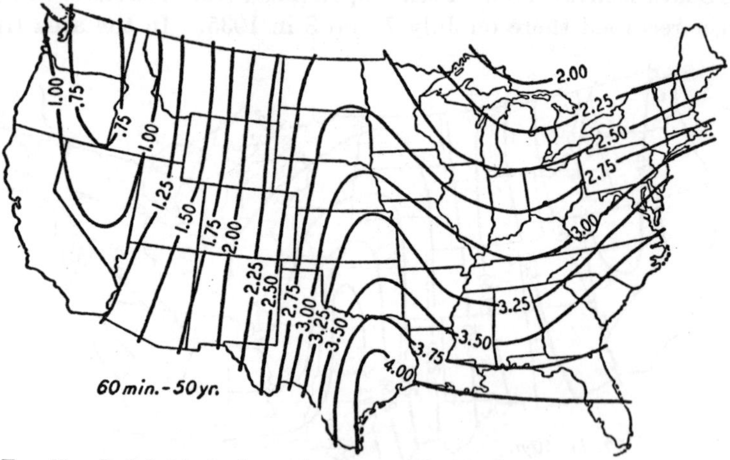

FIG. 31.—Rainfall in inches to be expected in 60 minutes once in 50 years.

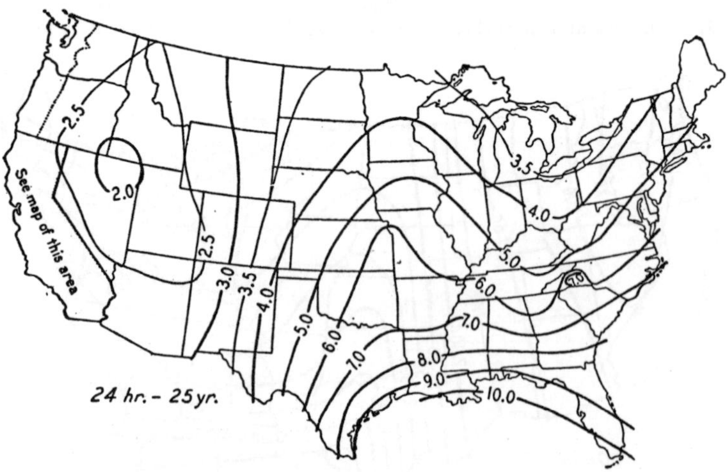

FIG. 32.—Rainfall in inches to be expected in 24 hours once in 25 years.

destruction was so great as to suggest that the actual rainfall was far heavier than that recorded at any of the stations by the official rain gauges.

The only official automatic weather-bureau rain gauge in the area of heaviest rainfall was located at the U. S. Weather Bureau station at Ithaca, N. Y. Previous records of severe thunderstorms show higher intensities for periods of 5 minutes to 2 hours than were recorded during the July, 1935, storm. The weather

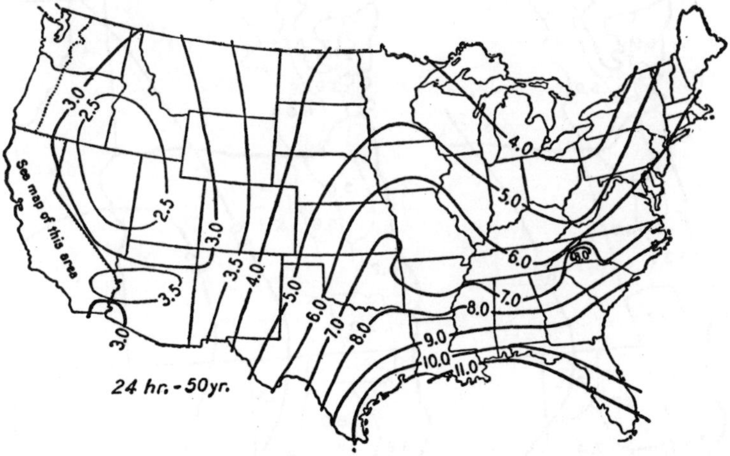

FIG. 33.—Rainfall in inches to be expected in 24 hours once in 50 years.

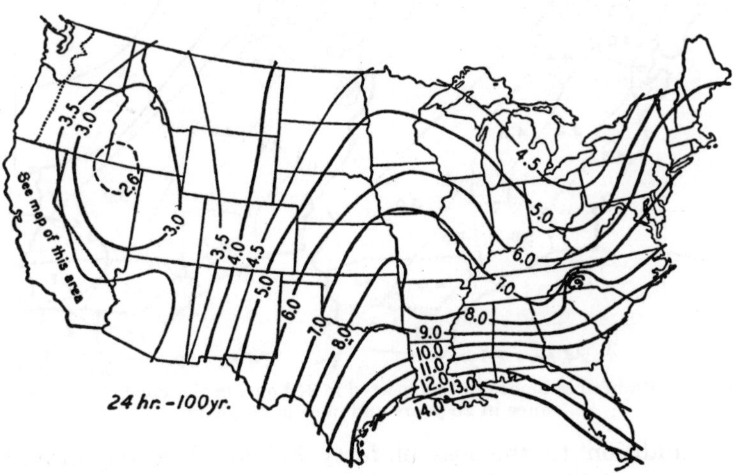

FIG. 34.—Rainfall in inches to be expected in 24 hours once in 100 years.

records have been kept at Ithaca for the past seventy-seven years. While the previous highest 24-hour record was 4.70 inches, 7.90 inches of rain fell between 4:00 p.m. on July 7 and 4:00 p.m.,

July 8, 1935, at Ithaca.[1] The severe floods resulted not from cloudburst showers but from the long-continued heavy excess of rainfall over the absorptive capacity of the soils of the region.

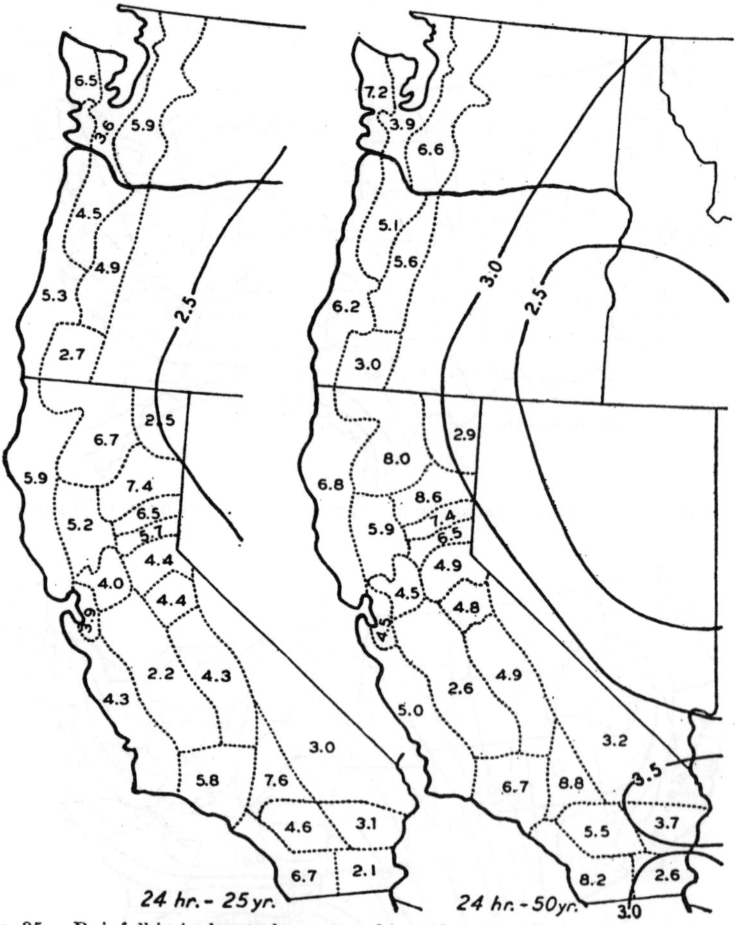

FIG. 35.—Rainfall in inches to be expected in 24 hours in the Pacific coast district once in 25 years and once in 50 years.

In addition to the loss of forty human lives the property damage was estimated at not less than 25 million dollars. Highways, utilities, railroads, state parks, and private homes in cities and villages and on farms appeared to be the heaviest sufferers.

[1] FISHER, J. C., Climatological Data, New York Section, *U. S. Weather Bur.*, July, 1935.

While this loss appears heavy, the loss in crops and vineyards was no less severe. The greatest damage of all, however, is represented by the loss of valuable surface soil and from gully formation on the upland farms. In addition to this damage, hundreds of acres of the best valley land was covered, and rendered almost wholly valueless, by stones and gravel deposited there by the flood waters. In Table 5 are given the records of the rainfall during that storm at central New York Weather Bureau stations.

TABLE 5.—RAINFALL AT STATIONS IN THE SOUTH-CENTRAL NEW YORK JULY, 1935, FLOOD ZONE*

Station	Rainfall, inches	Station	Rainfall, inches
Alfred	5.58	Cortland	11.15§
Haskinville	6.70	Geneva	1.41
Scio	3.85	Auburn	0.90
Andover	6.07†	Syracuse	3.30†
Angelica	5.08‡	Binghamton	1.02
Woodhull	1.55	Bainbridge	5.35‡
Hammondsport	6.10	Cooperstown	4.49
Dansville	4.69†	Honk Falls	6.10†
Hemlock	1.13	Morrisville	5.04†
Letchworth Park	1.81‡	Norwich	9.07†
Penn Yan	5.33†	Oneonta	6.71
Elmira	1.56‡	Delhi	8.52
Burdette	8.50	Sharon Springs	1.85
Ithaca	9.34§	Roxbury	4.15†

* Data supplied by J. C. Fisher, U. S. Weather Bureau, Ithaca, N. Y.
† July 7 and 8.
‡ July 8 and 9.
§ July 7, 8, and 9.

This rainfall came as a quick succession of thunderstorms, most of which actually occurred between the morning of July 7 and that of July 8. In other sections of the state the rainfall was very light on those days. Even so, the downpour covered an unusually large area. It was actually much heavier than would be expected once in one hundred years based on Yarnell's intensity-frequency charts for 24 hours. These charts would have suggested around 5 inches as the maximum for south-central New York. The previous maximum rainfall for 24 hours was 4.70 inches, which is in accord with the charts. It is to be hoped, however, that the 1935 24-hour rainfall record will stand for many years to come.

This storm has been treated in some detail because of its being representative of the extremely heavy and exceedingly unusual rains for the Northeast and because it is representative of the heavy rains of other sections.

Figure 36 shows the area on which between 5 and 10 inches of rain fell and that on which more than 10 inches fell during those

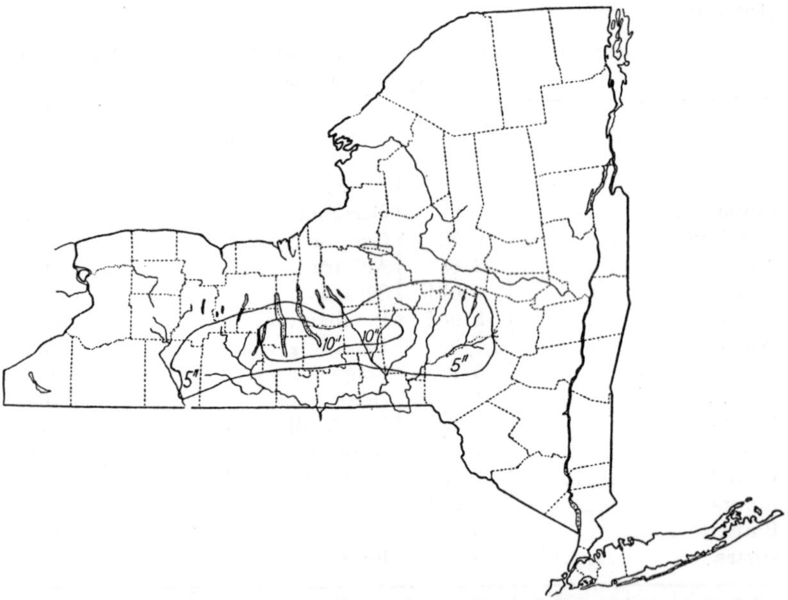

FIG. 36.—Areas in southern New York which received between 5 and 10 inches of rainfall in 4 days, but more than half of it fell within 24 hours. The area inside the 10-inch line received more than 10 inches of water.

four days July 6, 7, 8, and 9, 1935. In many places, however, most of the rain fell on the seventh and eighth.

Again it may be stated that man cannot, of course, influence the quantity of rain or the time that it may fall. Nevertheless, a knowledge of the quantity that may be expected should enable the farmer, the highway engineer, and the flood-control engineer, respectively, to plan his cropping system, the highway sluices, or the flood-protection works more wisely than any of them could do without fairly complete information on the expected rainfall intensities.

Runoff through Rivers.—The loss of rainwater through runoff varies greatly with the total rainfall, season, and type of precipitation and the soil, the agriculture, and the topography of the watershed. In Table 6 is given the runoff for a number of American rivers. An indefinite relationship may be expected between the percentage of runoff and soil erosion. The Illinois River is representative of the smaller of the large streams whose watersheds are comparatively level and which as a result will have slow runoff and relatively slight erosion in comparison to the total runoff.

TABLE 6.—RUNOFF THROUGH SOME AMERICAN RIVERS*

River basin	Years	Loss as drainage, per cent
Potomac	1886–1891	53.0 †
Connecticut	13	56.5 †
Savannah	1884–1891	48.9 †
Illinois (Greenleaf)	....	24.0
Kaskaskia (Illinois Geological Survey)	....	37.9

* MOSIER, J. G., and A. F. GUSTAFSON, Washing of Soils and Methods of Prevention, *Illinois Agr. Exp. Sta., Bull.* 207, p. 518, 1918.
† NEWELL, F. H., U. S. Geol. Survey.

Streams flowing through an area where much of the year's precipitation comes as heavy showers, whose watersheds are rather rugged, and whose soils are planted to cultivated crops have a relatively high percentage of runoff and relatively severe erosion.

For broad valleys and gentle slopes in open country it has been estimated that with a mean annual rainfall of 50 inches the average runoff will be about 25 inches, or 50 per cent; with a rainfall of 40 inches, a runoff of 15 inches, or 37.5 per cent; and with 30 inches of rainfall, a runoff of 8 inches, or 26.6 per cent.

In Table 7 are given the relative quantities of soil materials carried in solution and in suspension and the quantity of solids lost from each square mile of drainage area for a number of American streams.

It must be recognized that these data will vary widely with conditions for most streams. Ordinarily, a much higher pro-

TABLE 7.—MATERIAL CARRIED IN SOLUTION AND IN SUSPENSION BY AMERICAN RIVERS*

River	Location at which sample was collected	In solution, p.p.m.	In suspension, p.p.m.	Solids per square mile, tons
Mississippi	Minneapolis, Minn.	200	8	6
Mississippi	Jefferson Barracks, Mo.	206	964	250
Mississippi	Memphis, Tenn.	202	519	279
Missouri	Kansas City, Mo.	426	2,032	286
Ohio	Cincinnati, Ohio	120	230	363
Tennessee	Gilbertsville, Ky.	101	127	200
Arkansas	Little Rock, Ark.	630	748	145
Red	Shreveport, La.	561	870	244
Kennebec	Waterville, Me.	48	4	7
Hudson	Albany, N. Y.	107	17	30
Susquehanna	Williamsport, Pa.	74	18	30
Potomac	Great Falls, Md.	115	85	94
Roanoke	South Boston, Va.	71	264	348
Sautee	Camden, S. C.	73	214	303
Colorado	Austin, Tex.	231	351	26
Colorado	Yuma, Ariz.	710	5,354	387
Rio Grande	El Paso, Tex.	700	14,140	418
Sacramento	Sacramento, Calif.	121	60	86
St. Lawrence	Ogdensburg, N. Y.	134	Trace	1

* BENNETT, H. H., U. S. Dept. Agr., Graduate School, *Lecture* 1, Jan. 30, 1928.

portion of solids is found in suspension at flood stages as compared with the normal stage. The St. Lawrence is representative of streams that flow from a settling basin. The Great Lakes have removed the silt from the water flowing into them: consequently, their outlet, the St. Lawrence River, flows clear, essentially free of suspended soil material. Streams draining from large forest areas are in a similar condition. Little soil material reaches these forest streams, as it has never been started or has been filtered out by the forest litter. The Kennebec is representative of this condition. The Missouri is noted for the large proportion of sediment that it carries. At times when the Mississippi is flowing relatively clear the water below its junction with the Missouri is clear on the east side and distinctly muddy on the west side for miles below the mouth of the Missouri. The Missouri carries a much higher proportion of sediment than does the Mississippi on the average throughout the year.

Downpour Rainfall Produces Severe Erosion in Semiarid Regions.—In semiarid regions some of the soil is loose and fluffy silt loam, and some of it is loose, incoherent sandy material. Clay enough to bind the soil particles together is seldom present. The soil is usually low in organic matter owing to lack of sufficient rainfall to have produced luxuriant vegetation in the past. Trees and shrubs are absent or very scattered, and the grass is often so thin as to give little protection to the soil. Moreover, much of the inadequate supply of water comes as heavy, short-duration downpours which are often accompanied by hail. The beating of the raindrops and hailstones stirs the soil into a thick mud which tends to seal the pores and prevents absorption. This condition leads to excessive runoff, except on the coarser soils, but even on these the runoff is heavy. Once water starts flowing over the surface of these loose soils, severe erosion is underway. The extent of erosion is limited only by quantity of rainfall.

INFLUENCE OF LATITUDE

In the northern part of the United States the soil is normally frozen for from 3 to 4 months or more out of the year. In extreme cases, and especially at the higher altitudes, the period during which the soil is frozen is considerably longer than 4 months. To the southward the period during which the soil is frozen becomes shorter and shorter until along the Gulf Coast freezing of the soil is unknown.

While the soil is solidly frozen no washing occurs. Alternate thawing and freezing of heavy soils, particularly, permit the soil to slip or run down slopes. Sufficient water is usually present at that season to facilitate this movement. Rain or water from snow carries away considerable soil each spring under these conditions. However, if rapid thawing is accompanied by warm rains, the thawed part of the soil, especially if the land is bare, washes very readily. In fact, the soil in this semiliquid condition flows off, rather than washes away. If rain dilutes the semiliquid soil, all of the thawed part may be lost. Over a period of years this type of washing does occur rather frequently and is a source of serious loss of soil.

In the southern states, on the other hand, little or no protection is afforded by frost, and erosion is a year-round threat. Land protected by forest or by a thick vegetative cover is essentially as safe from erosion in one section of the country as in another. Land that produced such clean-tilled crops as cotton, corn, potatoes, peanuts, and vegetables the previous season, if left bare, is exposed to severe erosion throughout the winter. Intermediate between these extremes of frost protection against erosion are all the possible variations in severity of erosion. The erosion problem, consequently, may be more difficult of solution in the South than in the North.

Northern sections have another advantage over southern ones in that a share of the year's precipitation falls as snow. Erosion does not usually result directly from snowfall, as it often does from rainfall. Snow that falls on unfrozen soil thaws and is absorbed directly by the soil without runoff or erosion. Even if snow falls on frozen soil or if the soil freezes after the snow falls, there is usually less tendency for the water from snow to cause erosion than there is for the same quantity of water falling as rain. The reason for this is that water forms from snow at a rate slower than the average rate of rainfall and that the beating action of the raindrops is absent in the case of snowfall.

CHAPTER V

INFLUENCE OF SLOPE AND SOIL ON EROSION

Slope and soil, as well as precipitation and latitude, affect the rapidity and extent of erosion under varied conditions of land cover and soil utilization.

INFLUENCE OF SLOPE

Slope may be expressed in degrees or in per cent. A slope midway between perpendicular and horizontal is said to be one of 45 degrees. In practical agricultural work, slope is expressed in per cent. A drop of 10 feet, perpendicular, in 100 feet of horizontal distance is called a slope of 10 per cent and one of 25 feet in 100 feet horizontal is a slope of 25 per cent.[1]

The size of particles moved by streams varies from the smallest clay particles, which are carried in suspension, to large stones and boulders which are slid or rolled along on steeply sloping stream beds. The capacity of a stream for carrying soil material in suspension is limited and no more can be taken on unless the velocity is increased. The size of material which a stream can carry depends on the ratio of surface to mass of the soil particles and on the velocity of the stream.

Another point is worthy of consideration, since it helps in an understanding of the suspension of soil material in water. When

[1] The percentage of slope may be determined in several ways. (1) Take a light piece of wood, 100 inches in length, with a reasonably straight edge. Place a carpenter's or mason's level on the straightedge, one end of which is resting on the land whose percentage of slope is to be determined. Measure the distance from the soil to the lower side of the straightedge in inches when the level shows the straightedge to be horizontal. Since the straightedge is 100 inches long, the distance in inches from its outer end to the ground is the per cent of slope of the land. (2) An easier way of determining the percentage of slope is by means of a level of the Abney type. One sights at an object at the approximate level of one's eye on land whose percentage of slope is to be determined. The level is adjusted until it is at true horizontal. The reading is then made directly in percentage of slope or in degree of slope, depending on the calibration of the instrument.

a body is immersed in water it displaces its own volume of water whose specific gravity is unity. If soil particles which have a specific gravity of 2.5 are immersed in water, their effective specific gravity is reduced by that of the water displaced, or to 1.5. Furthermore, a film of water adheres to soil particles in suspension; in fact, the film acts as if it were a part of the soil grain itself. The weight of the adhering film is such a high proportion of the total weight of small particles, including the film, that their effective specific gravity is actually reduced to essentially that of water itself. And these facts explain how soil particles are kept in suspension by relatively slow-moving streams.

Table 8 is quoted by Geikie from Stevenson. These figures show that stream currents that appear to be extremely slow are capable of transporting materials of surprisingly large size.

TABLE 8.—SIZE OF MATERIAL MOVED BY STREAMS OF VARIED VELOCITY*

Velocity		Size of material transported
Inches per second	Miles per hour	
3	0.1700	Will just begin to work on fine clay
6	0.3400	Will lift fine sand
8	0.4545	Will lift sand as coarse as linseed
12	0.6819	Will sweep along fine gravel
24	1.3638	Will roll along rounded pebbles 1 inch in diameter
36	2.0450	Will sweep along slippery angular stones of the size of an egg

* GEIKIE, A., "Textbook of Geology," 3d ed., The Macmillan Company, New York, p. 380, 1893.

Steepness of Slope.—Water flows slowly over a gentle slope, say of 3 per cent, somewhat more rapidly over one of 10 per cent, and still more swiftly over a steeper slope, say one of 25 per cent. Other factors being similar, water flows faster the steeper the slope, of course. A rapid current cuts or erodes the soil surface over which it flows more rapidly than does a slow current. Increasing the velocity of a stream increases its cutting or eroding power to the square of its velocity; that is, doubling the velocity of the current increases the eroding power to four times that of the original velocity. A rapid current carries larger soil particles than does a slow one. Doubling the velocity increases the size of particle that the current can carry to the sixth power of its velocity. In other words, the current of doubled velocity can carry particles sixty-four times the size of those that the current carried at its original velocity. A stream that can carry particles 0.01 inch in diameter can, by having its velocity doubled, carry in suspension particles 0.64 inch in diameter. In the same way a rapid current carries more suspended material than does a slow current. According to Deacon, doubling the surface velocity

FIG. 37.—A steep slope on a drumlin in central New York. This Ontario loam is a valuable soil but it erodes actively on the steeper slopes.

increases the weight of material transported to the fifth power[1] or a little more, or it enables the current whose velocity is doubled to carry thirty-two times the quantity of soil material that the original current was capable of carrying.

The corresponding figures for trebling velocity are: nine times the original erosive power, ability to carry particles 729 times as large as could have been carried by the original velocity, and 243 times the quantity of material that could have been carried by the original velocity.

These figures seem incredible, but so, also, do the size and quantity of material that streams actually move down steep slopes. The greatly magnified power of swift currents as com-

[1] DEACON, G. F., Discussion on training of Rivers, *Minutes Proc. Inst. Civil Eng., London,* Vol. 118, p. 95, 1894.

pared with that of slow ones fully explains the work of streams at flood stage on steep slopes (Fig. 37).

Length of Slope.—The length of slope is an important element in soil erosion, especially on cultivated lands. A large quantity of water falls on a long slope. The water accumulates over the entire length of the slope unless the soil and land cover are such as to absorb rain as fast as it falls. If the rate of rainfall exceeds

FIG. 38.—A long erodible slope in northern Pennsylvania. Note the active gully in the foreground and the healed ones on the steeper slope.

the absorptive capacity of the soil under the existing land-cover conditions, the accumulation of water at the base of the slope often presents a difficult problem. The management of a short slope, on the other hand, is comparatively simple because the quantity of water collected is relatively small. Long slopes, unless very gentle, are usually accompanied by decided relief or large differences in elevation. The safe conduct of a large accumulation of water down a slope over a drop of several hundred feet in elevation is always attended with difficulties. Stabilized outlets over long distances on steep slopes are not easily found. Moreover, a drainage way that was perfectly stable for its normal quantity of water may become decidedly

FIG. 39.—A long slope in Virginia. This slope was all plowed for a clean-tilled crop in the spring of 1936. Destructive erosion may be the result.

FIG. 40.—Erosion caused by water from the concentration area above the steep slope. Central southern New York. This washing was caused by an ordinary heavy shower in June, 1936.

unstable if much additional water is turned into it. Under the latter conditions serious new gullying may develop, and, owing to possible greater accumulation as a result of control measures, the latter condition may be worse than the original unless due care is exercised. Long slopes in cultivated areas always present real problems (Figs. 38 and 39).

Concentration Area at Top of Slope.—Above long slopes on many types of geological and soil formations is a comparatively flat area, which, however, slopes sufficiently to permit drainage from it. Water accumulates on such land to a menacing extent whenever rain falls faster than the soil absorbs it. The water from such an accumulation area in nature is added to that originating on the slope itself. And it increases both the volume and the velocity of the water once it has passed down on to the steeper slope. Such additional water aggravates, immeasurably, the problem of controlling erosion on long slopes (Figs. 8 and 40).

INFLUENCE OF THE SOIL

In addition to precipitation, latitude, and slope of the land the soil itself is a large factor in erosion. It is quite fruitless, however, to question whether the soil is a more or less important factor in erosion than are precipitation and slope. Soils vary in their erosiveness or in their resistance to undesirable soil movement by water. In thinking of soils in relation to erosion two viewpoints need to be kept in mind: the rapidity of absorption of water by the soil and the ability of the soil to retain water or its water-holding capacity. The retention of water, however, has little direct bearing on resistance to erosion but very much to do with the possible production of a protective vegetative cover.

Size of Soil Particles.—The size of soil particles determines in a general way the size of the openings between soil particles in nature and thus influences directly the rapidity with which the soil takes up water. A coarse sandy soil, for example, takes up water very readily and permits little if any runoff from normal rainfall. Such a soil does not wash badly under average precipitation. But such a soil contains little fine material to bind it together and consequently may erode very, very easily if a stream of water is turned onto such a coarse soil. Coarse soils often do not hold sufficient water in droughty periods for good growth of crops or for the production of a vegetative cover.

At the other extreme are the fine-textured soils, such as clays and clay loams. The openings between actual particles in these heavy soils are so minute that water percolates into such soils with extreme slowness. In fact, percolation is so slow that a very large proportion of even ordinary showers is lost as runoff from heavy soils on moderate to steep slopes. Once the water has been taken up on heavy soils, however, it is held for the use of crops or for the production of a protective cover.

Silt loams, loams, and fine sandy loams are among the more desirable soils classes from the standpoint of absorption of rainfall and of retaining water for use by vegetation. A mixture of a wide range in size of soil particles is desirable in connection with avoidance, or control, of erosion.

Unusual uniformity in size of particles is characteristic of wind-laid material, such as the deep-loess deposits of the Mississippi and Missouri River valleys and also of soils laid down in lake or other quiet waters.

The reason for this condition is obvious upon consideration. The wind carried the finest particles far beyond the zone of deposition of the material that constitutes the deep-loess deposits. Particles much larger than those in the loess remained behind as sand dunes because they were too heavy to be borne very far in suspension in the air. This explains the marked uniformity in size of particles in deep loess. For these reasons loess does not contain any considerable proportion of clay to aid in binding it together. And this lack of binding material is undoubtedly a factor that contributes to the ease with which deep-loess deposits erode.

A somewhat similar situation obtains with respect to lake-laid soils. The coarser materials, such as sands, are deposited by streams as alluvial-fan or as delta material near the shore. Beyond that the silt is deposited, and farther from the shore the clays settle out and build up the clay deposits. Some mixing of silt and clay occurs, to be sure, as a result of variations in stream velocity and from wave action, particularly after the clay had been built up to a point where it was covered with only shallow water.

Raindrops falling on fine-textured lake-laid soils appear to be unusually effective in beating soil particles into suspension. And once in suspension, with the aid of the buoyant effect of water, these fine particles may be carried over indefinite distances by even slow-moving stream currents. It is notable that a lake-laid soil such as the Dunkirk erodes on very gradual slopes with extreme ease. In fact, slopes of from 3 to as low as 1 per cent require careful management under cultivation in order to avoid loss of surface soil.

In Table 9 is given the mechanical analysis of three samples of Miami silt loam from the Mississippi bluffs in Illinois east of St. Louis. The Hudson clay loam and Dutchess silt loam are from Dutchess County, New York.

TABLE 9.—MECHANICAL ANALYSES OF SOILS

Soil separate	Miami silt loam,* per cent			Hudson† clay loam, per cent	Dutchess† silt loam, per cent
	I	II	III		
Gravel (2 to 1 mm.)	0.06	0.52	0.02	0.0	4.0
Coarse sand (1 to 0.5 mm.)	0.34	1.00	0.36	2.1	7.4
Medium sand (0.5 to 0.25 mm.)	0.36	0.34	0.24	1.4	3.8
Fine sand (0.25 to 0.1 mm.)	0.78	0.46	0.58	5.2	8.7
Very fine sand (0.1 to 0.05 mm.)	2.80	3.34	3.60	6.4	4.6
Silt (0.05 to 0.005 mm.)	87.46	85.30	85.64	63.2	60.6
Clay (0.005 to 0.0001 mm.)	7.88	8.92	9.56	22.5	10.9

* Field Operations, *U. S. Dept. Agr., Bur. Soils*, p. 513, 1902. This was called fine sandy loam in the field.

† Field Operations, *U. S. Dept. Agr., Bur. Soils*, pp. 52, 64, 1907.

Attention is called to the uniformity of the Miami silt loams. These average 94.92 per cent of silt and clay of which 86.13 per cent is silt, according to the Bureau of Soils analyses. An appreciable percentage of a wider range of particles is found in the Hudson clay loam and Dutchess silt loam from New York.

Tilth and Porosity of Soils.—Good tilth or physical condition is highly desirable in all soils at all times. The maintenance of a goodly supply of organic matter is essential in all kinds of soils. Organic matter aids materially in keeping silt loams, clay loams, and clays in a granular condition. If the individual particles in these soils act separately, the spaces between them, called pores, are so minute that such soils take up water only with great difficulty and very slowly. In heavy soils, organic matter tends to cause the tiny particles to cling together in granules which are many times the size of the individual soil particles of which they are composed. The granules act very much as do sand particles of the same size. And a well-granulated soil possesses many of the good qualities of a sandy soil, without the less desirable ones such as droughtiness. The pore spaces between the granules are large enough to permit desirably rapid percolation of water under normal conditions even into a heavy soil.

If a large proportion of rainfall enters the soil, correspondingly less is lost as runoff, and erosion is simultaneously held in check. Unlike a coarse sand, however, a well-granulated heavy soil not only takes up but retains a large quantity of water for the benefit of crops and for the production of cover. In fact, building up and maintaining a fair supply of organic matter in soils is but one short step in restoring to soils the absorptive and retentive capacity that they possessed when first plowed from the original forest or from the prairie sod condition.

Depth and Water-holding Capacity of Soils.—The depth of the soil and its water-holding capacity have an obvious bearing on erosion. Deep, open, porous soils, especially those of medium texture, with a high proportion of pore space have a large capacity for water. The water taken up by the soil does not go off over the surface and thus cannot cause erosion. Soils vary greatly in depth and in their capacity for storing water (Fig. 41).

In connection with experimental work, Musgrave[1] used the term *infiltration capacity*. The infiltration capacity of a soil is the sum of the rate of percolation of water into the soil and the downward movement of water by capillarity. Determination of the infiltration capacity of a soil is useful in connection with consideration of the erosion-control measures needed on a particular soil under the varied conditions of rainfall, cropping system, and land cover. It must be borne in mind, however, that the infiltration capacity of soils varies with their organic-matter content and with their tilth. Looseness or compactness resulting from recent tillage or lack of it has a marked influence on the rate of infiltration of soils.

Influence of the Subsoil on Erosion.—The constitution and the condition of the surface soil, alone, do not determine the erodibility of a soil. One that has a loose, porous surface soil and

[1] MUSGRAVE, G. W., The Infiltration Capacity of Soils in Relation to the Control of Surface Runoff and Erosion, *Jour. Amer. Soc. Agron.*, Vol. 27, pp. 336–345, 1935.

FIG. 41.—Loss of rich top soil the result of cultivation. Shelby silt loam, similar to Fig. 1 in *Missouri Agr. Exp. Sta. Bull.* 349 by L. D. Baver, p. 7, 1935. (*Courtesy of U. S. Soil Conservation Service.*)

whose subsurface and subsoil are, also, open and of medium texture has a relatively high water-holding capacity. Such a soil of good depth will take up and hold a considerable quantity of water. Contrast this with a soil whose surface 6 to 8 inches is relatively loose and open but whose subsurface soil is a hard, compact, relatively tight or impervious stratum. This soil has a low water-holding capacity. Once the surface soil has become saturated, additional water falling on it or coming onto it, from whatever source, must pass off over the surface with consequent loss of soil from exposed slopes. The condition of the subsoil obviously is of great importance. In fact, in so far as the soil itself is concerned, the subsoil may often be the most important single factor bearing on its erodibility. The thickness of absorptive material over a tight substratum, or bedrock, determines very largely the quantity of water that a soil can absorb and incidentally the relative probability of destructive erosion.

In the field the relative erosiveness of a shallow-surfaced soil on a tight, compact substratum as compared with a deep, open, absorptive soil is very marked. Frost action causes erosion directly on shallow soils, and indirectly as well, through the heaving and the destruction of the vegetative cover. Such interference with some types of cover greatly encourages erosion in winter and early spring on sloping shallow soils.

Effect of Stones on Erosion.—One of the advantages of a mixture of sizes of soil material as compared with uniform or uniformly fine material is clearly shown by a comparison made at the Federal Soil Erosion Experimental Station near Ithaca, N. Y., and reported by Lamb.[1]

The soil under comparison is Lordstown stony silt loam, the slope 20 per cent, the length of plots 72.6 feet, and the rainfall from May 1 to Nov. 13, 1935, 23.96 inches. These plots were located outside the July 7 to 8, 1935, flood zone. The high rainfall intensities for 10-minute periods for the summer were as follows: for July, 0.60 inch; for August, 0.55 and 0.43; and for October, 0.46 inch. All stones larger than from 1½ to 2 inches in diameter were removed from plot A 7 but were left undisturbed on plot A 5. Cropping, fertilization, cultivation, and management were identical on the two plots. The removal of stones from plot A 7 was the only difference. The losses of water and soil are shown in Table 10.

TABLE 10.—EFFECT OF REMOVAL OF STONES ON LOSSES OF SOIL AND WATER

Plot	Crop	Conditions	Loss of water, per cent	Loss of soil, pounds per acre
A 5	Corn	No stones removed	11.09	8,580
A 7	Corn	Stones removed	18.92	16,192

Of course, it must be borne in mind that these data cover only a single season. Continuation of the comparison will probably alter them to some extent unless the stones are removed from plot A 7

[1] LAMB, JOHN, JR., *Soil Conservation News*, Ithaca, N. Y., p. 6, February, 1936.

from year to year. The differences were striking in 1935, the loss of water being 70 per cent greater with the stones removed, and the loss of soil being nearly twice as great with the stones removed as compared with the plot on which the stones were left undisturbed. It is perfectly clear that the stones afford a marked degree of protection to the land against loss of water and soil.

Special References

Technical Bulletins of the U. S. Department of Agriculcure. A Laboratory Study of the Field Percolation Rates of Soils, *Bull.* 232, 1931, by C. S. Slater and H. G. Byres.

Physical and Chemical Characteristics of the Soils from the Erosion Experiment Stations, *Bull.* 316, 1932; and Second Report on same, *Bull.* 430, 1934, by H. E. Middleton, C. S. Slater, and H. G. Byres.

CHAPTER VI

INFLUENCE OF NATURAL LAND COVER ON SOIL EROSION

Mention has already been made of the influence of land cover on soil erosion or the protection that vegetative cover affords to the soil against erosion. Two classes of natural cover are usually considered. These are forests and prairie grasses. While in nature these types of vegetative cover both accomplish essentially complete protection, they may well be considered separately because of some vital differences.

FOREST PROTECTION AGAINST EROSION

When the earliest settlers landed on these shores, they found the land covered with a dense forest growth. In fact, the entire eastern, southeastern, and central-northern parts of this country were clothed with forests (Fig. 1).

The soil under the trees was covered with litter which was composed of leaves, twigs, and fallen trees, all in various stages of decay. This litter protected the soil from erosion by insulating it against the action of falling raindrops. In the forest, falling raindrops do not strike the soil directly, as they do on cultivated soils. The force of the drops is broken as they fall on the needles and branches of the conifers or the leaves and branches of the deciduous trees. The water drips from the leaves or branches onto the litter or runs down the trunk of the trees directly into the soil. According to Lowdermilk,[1] the litter prevents raindrops from beating soil particles into suspension as occurs on unprotected soil surfaces. The suspended particles clog the pores and other openings and thus materially reduce percolation into the soil. Forest litter or a thick cover of grasses, therefore, prevents the sealing of the soil pores and permits normal, unhampered percolation of water into it.

[1] LOWDERMILK, W. C., Influence of Forest Litter on Run-off, Percolation, and Erosion, *Jour. Forestry*, Vol. 28, pp. 474–491, 1930.

Average, or normal, showers are completely absorbed by the soil. Some water passes off over the surface from heavy downpours, but it passes to the streams without carrying appreciable quantities of soil with it. The water entering the soil is added to and replenishes the ground-water supply which feeds springs, wells, and streams. Many old-time springs have dried up long since because of depletion of the ground water due to the clearing of the forests. The flow of streams has been greatly reduced at their low stage, and the peak of floods has been more than correspondingly raised by removal of the trees and by bringing the land under cultivation.

According to McGee,[1] the level of the ground water over an area of some 500,000 square miles in the eastern part of the United States has been lowered from 5 to 30 feet since the country was settled. It is obvious that such a drain must not be continued indefinitely and that from the standpoint of urban water supplies, as well as agriculture, action looking to the restoration of the supply of ground water is urgently needed.

The early settlers cut the forest for fuel and for sheltering themselves and their livestock. In order to produce food for their families and feed for their animals additional land had to be cleared and planted to crops. Since there was no market for lumber at that time, much excellent timber was burned to get rid of it. Somewhat later, millions of feet of the finest hard and soft woods was burned, and the ashes, or the potash leached from them, sold in Europe. In fact for a time potash was an important source of income.

As the seaboard cities developed, food for the populations had to be produced near by, since the oxcart was the principal means of transportation. As a result, all suitable land even though steeply sloping was cleared and planted to wheat and other food crops. There was no wanton destruction of forests. The clearing was done in order to use the land for crops or to obtain the income from the potash.

When the canals were opened and when some years later the railroads were built, much of the steepest, poorest, least accessible land suddenly dropped out of cultivation for obvious reasons.

[1] McGee, W. J., Soil Erosion, *U. S. Dept. Agr., Bur. Soils, Bull.* 71, p. 17, 1911.

Forests Retard Runoff.—The reason for the differences between reservoir conditions in the Adirondacks and those at Ithaca or in the Piedmont and in the Southwest is not difficult to find (Chap. III). Forests cover the watersheds of the Adirondack streams. The soil is protected by the trees and the litter covering the forest floor. The forest soil in its natural condition is loose and open and absorbs water readily. The water which soaks into the soil first of all supplies moisture for vegetation. The bulk of it

FIG. 42.—Water flowing from a forest 15 hours after a heavy rain. Water was flowing from this forested area about 15 hours after a long heavy rain had ceased. No water was running from pastures or cultivated fields at that time. The forest had delayed the flood water in reaching the streams by several hours and thus reduced the crest of the flood. Photograph taken in eastern New York.

becomes temporarily a ground-water supply which reappears later as springs or goes directly into streams. By thus holding back the rain water, forests reduce the peaks of high water and tend to maintain stream flow at a moderate level between rains. In other words, forests aid greatly in regulating the flow of streams (Fig. 42).

In spite of the absorptive capacity of forest soils some runoff of water occurs following exceptionally heavy rains. However, runoff water from large timbered areas is clear and carries very, very little sediment. As a consequence, reservoirs fed by forest streams do not silt up rapidly. If, however, a considerable percentage of the area supplying water to a reservoir is planted to intertilled crops, silt is washed from the higher land and deposited in the reservoirs, thus rapidly destroying their effectiveness.

In Table 11 is given a five-year summary of the effects of different types of land cover on losses of soil and water at Guthrie, Okla.

TABLE 11.—EFFECTS OF DIFFERENT TYPES OF LAND COVER ON LOSSES OF SOIL AND WATER, ON VERNON FINE SANDY LOAM, RED PLAINS SOIL EROSION STATION, GUTHRIE, OKLA.,* 1930–1934
(Average rainfall, control plots, 33.39 inches; woods plots, 33 inches)

Land cover	Slope, per cent	Loss of water, per cent	Loss of soil per acre, tons
Cultivated crop (cotton)	7.7	15.4	25.47
Small grain (wheat in rotation)	7.7	14.5	1.72
Hay (sweet clover in rotation)	7.7	8.3	0.62
Woods (burned off)	5.2	5.4	0.19
Bermuda sod (clipped)	7.7	1.4	0.04
Woods (virgin)	5.2	0.2	0.016

* LEWIS, H. G., and H. S. RIESBOL, Outline of Investigations and Summary of Results: 1930–1935, Red Plains Soil Erosion Exp. Sta., Guthrie, Okla., *U. S. Dept. Agr.*, SCS-AP-6, p. 4, 1936.

Particular attention is called to the fact that virgin forest lost only 0.2 per cent of the rainfall and only 0.016 ton, or 32 pounds, of soil an acre a year during this five-year period. On this very moderate slope, 5.2 per cent, the burned-off woods, also, lost very little soil. The loss on the cotton land, however, is suggestive of what may be expected on cultivated lands in that area. More than one-seventh of the year's rainfall was lost from the cotton plot, and 25.47 tons of soil was washed from the cotton land.

These figures explain why reservoirs in forested areas continue in service for many years and also why reservoirs in hilly areas growing cultivated crops silt up so rapidly.

Table 12 gives the soil and water losses for one year from wooded land as compared with various cropping and tillage conditions on Cecil sandy clay loam, at the University of Georgia.

The most important point for consideration at this moment is the effect of the protection afforded by the trees. The loss of

TABLE 12.—SOIL AND WATER LOSSES IN GEORGIA, JULY 1, 1933, TO JUNE 30, 1934*
(Cecil sandy clay loam, rainfall 41.85 inches)

Plot	Crop, soil conditions, and tillage	Water lost, inches	Soil lost, pounds per acre
A	Wooded area (trees cover the ground)	0.08	115
B	Badly eroded (subsoil exposed)	18.61	112,320
C	Continuous cover crop (cowpeas-Austrian peas)	1.48	1,518
D	Tilled (3 inches after rains if dry enough)	5.62	14,205
E	Deep tilled (6 inches after rains if dry enough)	7.14	16,037
G	Fallow land (tilled 8 inches deep, left smooth)	19.97	159,747

* COLLINS, W. C., Austrian Winter Peas and the Vetches, *Univ. of Georgia, Bull.*, 10*b*, p. 5, 1935.

water was wholly negligible and of soil exceedingly slight. The badly eroded and the fallow land lost nearly half of the rainfall and 112,320 and 159,747 pounds of soil to the acre, respectively. These losses of both soil and water were enormous. Such loss of soil must lead rapidly to complete land ruin. These data cover only a single year, but it is notable that the trees had an effect very similar to that observed at Guthrie, Okla., reported in Table 11.

In Table 13 is shown the effectiveness of a forest cover in holding rainfall in Wisconsin.

TABLE 13.—MAXIMUM RUNOFF RATES IN UNGLACIATED AREA OF WISCONSIN, SUMMER OF 1929*

Crop or land cover	Number of measurements	Average maximum runoff, per cent	Average steepness of slopes, per cent
Forests	12	2.8	36.3
Native pasture grasses	5	7.2	34.3
Cultivated hay	3	17.7	15.3
Small grain and fallow	6	25.6	15.0
Corn	6	25.7	15.2
Seeded pasture, mostly bluegrass	7	26.7	31.6

* COHEE, MELVILLE H., Erosion and Land Utilization in the Driftless Area of Wisconsin, *Univ. Wisconsin, Stencil Circ.* 153, p. 3, 1934.

The maximum forest runoff was 2.8 per cent on a 36.2-per-cent slope. In comparison, cultivated grass (hay), which is usually regarded as relatively effective in holding rainfall, lost 17.7 per cent on a 15.3-per-cent slope. The forested slope was two and one-half times as steep as the meadow slope. This steeper slope might have been expected to lead to almost three times as much runoff from the forest land. The meadow runoff was six times as great as from the forest. If this be combined with the disadvantage of the steeper slope in the forest, it is clear that the forest was between fifteen and eighteen times as effective in holding back water as was the meadow under Wisconsin conditions in the driftless area.

At Tyler, Tex., according to Hendrickson and Baird, forested Kirvin soil having a slope of 12.5 per cent lost 0.7 per cent of the 42.19 inches of rainfall as an average for the three years 1932–1934 and only 0.083 ton, or 166 pounds, of soil.[1] A loss of 166 pounds of soil is, indeed, negligible. At that rate of erosion a period of 1,200 years is required for the removal of the surface 7 inches of soil.

In nature the forests held the soil in place, and the earliest settlers had no erosion problem at first. During a few years of cultivation, however, the forest litter and the rootlets decayed and disappeared from the soil. And with them went the last of the perfect protection that nature had provided so well over the long years. As more and more of the absorptive surface soil was lost, erosion became easier and easier, until finally, owing to the keen competition of the more easily tilled lands of the Midwest, thousands of acres of eastern hill lands have been abandoned.

Man cut the forest for its own products or for those of the land when cleared, and soil erosion followed. In a large measure man can in time reclothe the steeper and less productive hills with productive forests.

PRAIRIE-GRASS PROTECTION AGAINST EROSION

Prairie grasses in the Midwest and on the Great Plains naturally protected the soil quite as effectively as did the trees in the forested areas. The beating raindrops could no more reach the

[1] HENDRICKSON, B. H., and R. W. BAIRD, Summary of Principal Results Soil Erosion Experiments, Tyler, Tex., *U. S. Dept. Agr., Soil Conservation Service*. SCS-EP-3, p. 15, 1935.

soil under the prairie grasses because of the protecting layer of organic matter than under the trees. The blades and stems of the grasses conducted the water into the soil much as did the leaves, branches, and trunks of the trees. Under the grasses as under the trees the water soaked into the soil to feed springs and streams in a steady regular manner. Unabsorbed or excess water from very heavy thundershowers passed off perfectly clear to the streams or swamps much as it did from the forested hills of the East. The native prairie grass with its mass of fibrous roots gave perfect protection to the soil on which it grew for many centuries.

Some years ago a sample of prairie sod was taken from a virgin area near Urbana, Ill.[1] The soil was a good brown silt loam. The top growth was cut off, and the soil was washed from the roots in the surface 6⅔ inches. The weight of roots was equivalent to 27,000 pounds an acre, which is truly an immense quantity of live roots. These grass roots extend to depths varying from 8 to 20 inches. It is little wonder that the virgin prairie sod had protected the soil so completely through the ages, and this in spite of the fact that prairie fires occasionally burned off the aboveground part of the grass.

As the tops fell down and as some of the roots died and all of this material decayed or was carried into the soil by earthworms and rodents, the organic-matter content of the soil was slowly built up. According to the analyses of 302 samples of rolling and flat prairie soils, these contained on the average 4.53 per cent of organic matter in the surface 6⅔ inches and 2.8 per cent in the stratum 6⅔ to 20 inches below the surface. The corresponding figures for 164 samples of rolling-timber soils were 1.93 per cent in the surface and 0.77 per cent in the subsurface soil. Both the prairie and the timber soils had been under cultivation for approximately the same length of time.

Not only did the organic matter from both forest and prairie vegetation protect the soil from the mechanical beating by raindrops, but upon decay it supplied the soil with humus as well as active organic matter. This kept the soil loose, friable, open—in fact, in good tilth. Being in good tilth, the soil absorbed water readily and thus reduced the runoff. On bare slopes the tend-

[1] Mosier, J. G., and A. F. Gustafson, "Soil Physics and Management," pp. 144–145 J. B. Lippincott, Philadelphia, 1917.

ency is for runoff water to flow rather rapidly, particularly after its volume has increased toward the lower part of the slope.

The organic matter of the forest and that of the prairie together with the stems of the grass check the flow of water on sloping land. Checking the flow of water encourages its absorption by the soil. This comes about mainly since the water is held in contact with the open soil for a longer period. This gives both gravity and capillarity more opportunity to do their work. Absorption is increased thereby, and runoff is reduced. As runoff is diminished on bare soils, erosion is reduced. Moreover, as absorption of water is increased, the ground-water supply is replenished, which, also, is universally desirable.

A few years after the native sod was plowed, the more active organic matter had decayed and slowly disappeared from the soil. Rain waters quickly found their way into the swamps and streams. Later the flat lands were tiled, and the swamps drained. This rushed the water into the streams much sooner after rains than occurred in nature or before the settlers plowed and drained the prairies. The stream channels were not large enough to carry the water forced into them by the accelerated drainage. Floods in the broad valleys of the Midwest were the result.

On the prairies, as in the forested areas, man can do something toward restoring the natural conditions by seeding a larger share of the more rolling prairie lands of the Midwest to cultivated grasses and legumes for hay and pasture. Even if it were possible and feasible to restore the native grasses it would be unnecessary in that area because the more productive cultivated grasses and legumes are relatively effective in the control of erosion. On the Great Plains, however, such native grasses as blue grama, buffalo grass, slender wheat grass, and little bluestem are exceptionally effective in controlling erosion of all kinds. In addition, lespedeza is of great service in certain areas.

CHAPTER VII

EFFECTS OF CROPS AND TILLAGE ON EROSION

The ordinary crops grown in this country fall into two groups with respect to their influence on soil erosion. These groups are the *close-growing* crops which retard erosion and the *clean-tilled* crops which encourage erosion.

CLOSE-GROWING CROPS

Close-growing crops are those which make a close cover over the soil and which afford real protection to the soil against erosion. These crops are seeded *broadcast* or *solid;* that is, with a grain drill all holes open and are called close-growing crops. They do grow *close,* because there is but relatively little space between the plants.

Hay Crops.—Not all close-growing crops are equally effective in preventing or in checking erosion. Those which have a fine, thick, matted root system are most effective in holding soil against erosion. What is needed, in fact, is a blanket of vegetation, a complete cover, which prevents raindrops from striking and churning the surface soil into mud, because the mud is diluted by further rain and carried off the field where nature placed it.

Grasses and Grass-legume Mixtures.—Timothy (*Phleum pratense*), Kentucky bluegrass (*Poa pratensis*), orchard grass (*Dactylis glomerata*) seeded thickly, and redtop (*Agrostis alba*) when making a thrifty growth are exceptionally resistant to erosion in the northern states. These grasses are effective in essentially the order stated (Figs. 43 and 44).

A mixture of legumes and grasses makes a denser sward than does grass grown alone. Moreover, mixing legumes with grasses enables the grasses to continue productive over a longer period than if the grasses are grown in the absence of legumes.

The lespedezas seeded alone serve in this way in the region to which they are well adapted. Some of the lespedezas thrive not only in the southern states but to the northward as far as central Illinois. In other words, the areas in which these legumes thrive overlap to a considerable extent. In the southern states Johnson grass (*Sorghum halepense*) is a good hay plant which resists erosion to excellent advantage (Fig. 45).

In the West the grama grass, especially blue grama (*Bouteloua gracilis*), and the buffalo grass (*Buchloë dactyloides*), both wild ones, are good pasture grasses, as is, also, western wheat grass

FIG. 43.—Grass roots check erosion on a highway. This sod of rough-stalked meadow grass (*Poa trivialis*) made a desperate fight against flood waters and was largely successful in holding the soil along the road across this valley. Two bridges were swept away. Photograph taken in central New York.

(*Agropyron smithii*). These grasses are effective in holding the soil against both wind and water erosion.

Clovers.—In the northern states the clovers are usually seeded with timothy or with timothy and other grasses. Mixing clovers and grasses is desirable not only from the standpoint of producing hay of high protein content but, also, from that of erosion control. Both medium red (*Trifolium pratense*) and alsike (*T. hybridum*) are usually seeded with grasses. So, also, is mammoth red clover. Each of these clovers has its own specific soil adaptations and, of course, gives the best account of itself when grown on soil to which it is well adapted. Clover-grass seed mixtures produce more feed and, incidentally, control erosion more effectively than do grasses grown alone without clover. In the northern states to which wild white clover (*Trifolium repens* var.) is adapted, it is highly desirable to grow it along with these grasses. Not only does this clover improve the feeding value of the hay, but it supplies nitrogen to the grasses. This effect of legumes

FIG. 44.—Grass roots check erosion on fields. The timothy was plowed in the spring. Approximately 2 inches of rain fell in 30 minutes on June 22, 1935. The holding effect of the grass roots is still very evident even though the grass was plowed under about two months previous to the taking of this picture. Slope 10 per cent. Photograph taken in southern New York.

explains why legume-grass mixtures produce more feed and control erosion to better advantage than do most grasses as ordinarily grown alone.

Great need exists for a perennial or a long-lived red clover in this country. To be of maximum service such a strain must be winter-hardy in the northern states. A perennial red clover would aid very greatly, indeed, in maintaining good thick swards of meadow plants for the protection of erosive sloping lands.

Alfalfa.—Alfalfa is a valuable crop for growing on erosive slopes. It has one great advantage over grasses. Being a legume it can obtain its own nitrogen from the air if other conditions are favorable. Its ability to produce large yields of hay of high feeding value and the relatively good market demand for it are decidedly in its favor. However, alfalfa has one distinct handicap, its narrow soil requirements. Its high lime need is

FIG. 45.—Lespedeza checks erosion in North Carolina. This lespedeza is growing on a "galled" spot from which the surface soil has been washed away. Lespedeza is one of the best crops to grow on such areas in the region to which it is adapted. (*Courtesy of U. S. Soil Conservation Service, North Carolina.*)

another; but since its lime requirements can be met, that handicap may be easily removed.

In its soil requirements alfalfa is discriminating. To be suitable for alfalfa a soil must be deep, open, and well drained. Since many sloping soils do not possess suitable drainage, the crop is somewhat limited in its usefulness in erosion-control work. Owing to its superior qualities, however, alfalfa may be used to advantage in the area adapted for its growth and on the soils well suited to it.

Alfalfa grows in bunches and does not produce a thick uniform soil cover. Some bare soil may be exposed to washing between the plants. Thicker-than-normal seeding will aid in developing a more uniform distribution of plants, but, even so, the bunch habit of growth must be recognized.

For market, pure alfalfa without any admixture of grass is desirable. For feeding on the farm, however, a mixture of alfalfa and grasses is preferable for many purposes. Moreover, evidence is at hand that alfalfa is longer lived when grown with grasses than if grown alone. If seeded together, the second growth is usually essentially clear alfalfa. For erosion control, furthermore, the addition of grass to seedings of alfalfa is desirable. The mixture makes a closer sward, a more complete cover, and gives much better protection against erosion than does alfalfa seeded alone. Frequently, of course, grasses in time come into stands of alfalfa, but the close sward desired is obtained sooner and with greater certainty if the proper seed mixture is sown (Fig. 46).

Fig. 46.—Alfalfa and grass hold the soil against erosion. This old alfalfa, along with the grass which has come into it, holds erosion of the soil to a minimum on the steep slopes of these drumlins in north central New York.

Annual Hay Crops—Millets and Sudan Grass.—Among the annual grass-hay crops are the millets and sudan grass (*Sorghum vulgare* var. *sudanense*). For many reasons these crops are limited as to acreage. During their period of active growth, from the time when they have reached a height of 4 to 6 inches to maturity, these grasses afford reasonably good protection to the soil, and the stubble holds the soil fairly well during the winter and early spring following.

The difficulty with them is that these crops are not seeded until late in the season. Moreover, they usually follow an intertilled crop. Consequently, the soil is essentially devoid of crop protection from the time when the intertilled crop is harvested through the fall, winter, and, spring—a period of six or seven months. Much erosion may take place during this long period of exposure.

FIG. 47.—Winter barley in Missouri, a close-growing crop. The small grains control erosion during the period of their growth and until the seeding in them is well established. (*Courtesy of W. C. Etheridge, Missouri Agr. Exp. Sta.*)

Small Grains.—The small grains are of two classes: the fall-seeded grains, represented by wheat and rye; and the spring-sown grains, represented by oats and barley.

Fall Grains.—The fall grains occupy the land over a period of from seven to nine months depending on the latitude in which they are grown. In Texas or Oklahoma wheat seeded in late October or early November is ready to harvest early in June, a period of seven months or slightly more. In New York, on the other hand, fall wheat occupies the land for nearly nine months.

The relatively long time that these crops are on the land is favorable for the control of erosion.

After wheat and rye have made a top growth of 3 inches they afford considerable protection, partly because by that time they have a strongly developed root system. And the mass of roots, particularly of rye, hold the soil and check erosion to a very marked degree. This protection by the crops is continued throughout the life of the plants and by the roots and stubble after the crops are harvested. When, as is usually the case, clover or alfalfa and grasses or other hay plants are seeded in the

FIG. 48.—Buckwheat nonresistant to erosion. This buckwheat field was badly gullied by one heavy rain. The water which collected in the buckwheat caused gully formation across the potato rows, the stones being carried into the pasture at the base of the slope.

fall grain, the protection may be continued through several additional years (Fig. 47).

Spring Grains.—The spring grains act in exactly the same way as do the fall grains. They are usually seeded after a cultivated crop such as corn, cabbage, beans, or potatoes, grown the previous year. Soil washing may take place following the preceding clean-tilled crop, as happens preceding millet or sudan grass because conditions are similar. Being seeded early in the spring, oats or barley control erosion to good advantage. In a section such as south-central New York, oats and barley are seeded early enough to become well established before the summer rains come on. Oats seeded on rather steep slopes on the Cohocton River, New York, U. S. Soil Conservation Service demonstration project in the spring of 1935 controlled erosion in a highly satisfactory manner even through the heaviest rainfall on record. After harvest the stubble and the new seeding control erosion very well, in fact precisely as do the fall grains.

Buckwheat.—Buckwheat (*Fagopyrum esculentum*) as usually grown in the northeastern states does not control erosion. The land is plowed and worked down somewhat, and the crop seeded in June or early July in the midst of the heavy thundershower season. The soil is loose. Given sufficient moisture the plants come up quickly, but they develop only a very limited root system. Consequently, the torrential rains of that season of the year bring about severe erosion of both the sheet and the gully types in buckwheat fields (Fig. 48).

Furthermore, after buckwheat is harvested in the late autumn, the stubble falls over, and the soil is exposed almost as completely as it is following late potatoes. Severe late-fall and early-spring washing follows buckwheat. For these reasons, it is not recommended, particularly for steep erosive soils. Rye is sometimes seeded with buckwheat. The rye makes little growth during the summer and does not interfere with the growth of the buckwheat. When cool weather comes after the buckwheat is harvested the rye makes some growth. Even a meager growth aids materially in the control of erosion throughout the winter and the following spring.

Pastures.—A good, thick, thrifty pasture sward controls erosion almost as completely as does virgin prairie grass or forest. A well-adapted perennial, or long-lived legume, is a prime requisite for thoroughly good pastures. An annual or a biennial legume which reseeds itself will serve the same purpose as a perennial one reasonably well.

A thick pasture sward intercepts the raindrops. They do not strike the soil directly and consequently do not beat the soil into suspension to be lost in the runoff water. Furthermore, the aboveground growth checks the flow of water and thus encourages its absorption by the soil, and the roots hold the soil against erosion.

A weak, thin sward, on the other hand, permits erosion to go on almost unchecked. Such a thin sward exposes a rather high proportion of the soil to the beating action of raindrops, which results in the formation of a thin mud. As this runs downhill it seals the soil pores by filling them with fine particles and renders conditions ideal for heavy runoff. On such pasture land rather severe erosion takes place, and gullying eventually follows.

It is more difficult to maintain a thick erosion-resistant sward on south and west exposures than on the cooler east and north slopes.

Annual Legumes—Cowpeas, Soybeans, and Cannery Peas.—Cowpeas and cannery peas are usually drilled solid, that is, with all drill holes open. Soybeans are seeded in this way when grown

Fig. 49.—Potatoes a clean-tilled crop. Some washing took place on this gentle slope after a light shower. Photographed in southern New York.

for hay, as are also some of the newly developed soybean varieties grown for grain. These crops do not develop fine fibrous root systems, and they do not control erosion nearly so well as do the grasses.

CLEAN-TILLED CROPS

The crops which are planted in rows far enough apart for intertillage and which are cultivated are called *clean-tilled* crops.

Cotton, Corn, Tobacco, Beans, Cabbage, Potatoes, Sorghums, Soybeans.—Cotton, corn, tobacco, dry and cannery beans, cabbage, potatoes, sorghum, soybeans for seed or for grain, kaffir and milo, and ordinary vegetables are rated as clean-tilled crops. Several of the crops usually grown as clean-tilled ones are sometimes seeded as close-growing crops. Corn, sorghum, and soybeans are sometimes sown broadcast or drilled solid and, under these conditions, control erosion much better than they do when they are clean tilled.

In the case of clean-tilled crops most of the soil is exposed to the action of the rain during the entire growing season and sometimes throughout the succeeding fall, winter, and early spring as well. Frequent tillage keeps the soil loose, a condition in which it is unusually susceptible to erosion. Some clean-tilled crops are more conducive to erosion than others. In the late-potato areas, for example, the potato crop encourages erosion markedly. This results from clean tillage throughout the season and from the fact that the harvesting operations leave the soil loose, smooth, and bare. Under such conditions steeply sloping potato lands undergo severe erosion (Fig. 49).

ROTATION OF CROPS

The beneficial effects of practicing rotation of crops including at least one legume, as compared with growing nonlegume crops continuously, may be stated briefly without discussion. These benefits are (1) better distribution of labor; (2) better control of insects, plant diseases, and weeds; (3) help in maintaining the organic-matter content of the soil; and (4) help in producing larger crop yields. The benefits from following a regular rotation of crops are obtained in addition to those derived from the application of farm manure, lime, and fertilizers.

The addition of organic matter to the soil by means of rotation of crops and the beneficial effects of organic matter on tilth are of special importance in the development of soil-erosion control programs. For this there are two reasons. The granules in soils of good tilth do not wash away so readily, because of their relatively large size, as do individual soil particles. And well-granulated soils absorb water more readily than do those which are not in good tilth. In a desirable rotation, especially for hilly lands, intertilled crops should occupy the soil only a relatively small proportion of the time. Not more than one such crop in four years is desirable on moderately erosive land, one in five years on somewhat more erosive soils, and as nearly as possible *permanent* hay or pasture on rather erosive lands which are required to produce the hay needed on the farm or for sale.

FRUIT CROPS

Fruit growers are confronted with a variety of soil-erosion problems depending on the slope of the fruit-producing lands, the erosivity of the soil, and the type of cover that may be produced and yet not interfere with the production of fruit. In the case of tree fruits, grapevines, and, in a lesser measure, the bush fruits it is out of the question to clear the land, owing to expense of reestablishing them, and start new, as may be done with the ordinary feed and vegetable crops.

Tree Fruits.—Apples, peaches, pears, cherries, nectarines, oranges, lemons, and grapefruit fall in the class of tree fruits. These trees have large root systems which help in a measure to hold the soil in place. Clean cultivation on sloping land, however, leads to erosion. In places so much soil has been removed by erosion from between the rows that the trees stand on a ridge, more or less high and dry. Many of the feeding roots have been exposed and killed, and the trees deprived of water and plant nutrients to such an extent that many trees eventually die. Others may survive for some years but have been so weakened as to be unable to produce fruit or to make a thrifty growth.

Vineyards.—In the eastern states, vineyards present a special erosion problem, since grapes are so generally clean cultivated. Grapes, being sensitive to frost, must be grown in places protected from frost, usually adjacent to good-sized bodies of water. In areas of marked relief, grapes are likely to be grown on relatively steep slopes facing the body of water.

In the Finger Lakes region of New York the slopes are generally steep. Along Lake Erie and Lake Ontario some of the protected zone is steeply sloping, but some of it is relatively flat. The flat land, however, is lake-laid material and consequently subject to erosion on very mild slopes. Such land with a slope of 5 per cent requires special protection against erosion.

Bush Fruits.—Bush fruits are grown on a variety of soils, some erosive and some nonerosive, and on different degrees of slope. Cultivation of the bush fruits is common practice. Consequently, the attendant erosion problem is similar to but usually much less difficult than that in steeply sloping vineyards, such as those bordering the Finger Lakes in New York.

CHAPTER VIII

MAINTENANCE OF ORGANIC MATTER; FERTILIZATION AND ROTATION OF CROPS

In the foregoing pages erosion conditions in the United States have been described. The various kinds of erosion have been discussed, and its results on the lands of the country have been presented. The effects of precipitation, latitude, slope of land, and of the soil itself on erosion have been discussed along with the factors influencing erosion which are in a measure under man's control. Among these were included different types of vegetative cover with which man may clothe the earth and the relationship of different types of crops to soil erosion.

Practical measures for the control of soil erosion may now be considered. In a broad general way the most important single factor in preventing erosion is the production and maintenance of a complete vegetative cover for the soil. Nature covered both forest and prairie soils with vegetation long before the white settler arrived in this country. In fact, it is mainly by interpreting, adapting, and applying nature's ways to present-day conditions that erosion-control methods have been developed. Nature's methods point to the growing of trees, shrubs, grasses, legumes, and small grains for the control of soil erosion. Close-growing crops which provide the maximum of protection against the beating action of raindrops and which by means of their innumerable blades, leaves, and stems check the flow of water on slopes are most effective. Vegetative protection is equally effective in the control of both water and wind erosion. One of the underlying principles of erosion control is the protection of the soil against direct action of either wind or rain.

The maintenance of the organic-matter content of the soil, fertilization and liming for the production of good crop yields and an adequate vegetative cover for the soil, contour tillage and seeding, and contour strip cropping are essential in any complete soil-erosion control program under normal conditions. For control of erosion under abnormally high-intensity rainfall, however, the use of interception or diversion ditches and terraces, in addition, is of utmost importance in many sections of the country. In fact, reasonably complete control of erosion, especially under the less favorable conditions, can be hoped for only when all of the known control measures are applied intelligently to the entire soil-erosion problem.

The control measures for erosion by water, by wind, and by waves are considered separately.

MAINTENANCE OF ORGANIC MATTER

Organic matter, as previously stated, aids in erosion control by bringing about highly desirable granulation of the soil. Granulation increases percolation into the soil and absorption of rainwater by the soil. The result is less runoff and consequently less erosion. In addition to its granulating effect, organic matter acts as a sponge, taking up and holding water for the use of crops. And most crops in many sections under average conditions could use to advantage all of the water that could be saved by ordinary additions of organic matter to the soil.

Southern soils lose organic matter more rapidly than do northern ones. This loss results from the longer period of temperatures sufficiently high for bacterial action or for the decay of organic matter in the South as compared with that in the North. Moreover, in parts of the South an exceedingly high proportion of the soil is plowed and planted to clean-cultivated crops each year, a cropping condition under which organic matter is rapidly destroyed.

Another point worthy of consideration is that southern soils originally were lower in active organic matter than were northern soils of the same general texture. This condition is well illustrated in Illinois where many soil analyses are available for comparison. The deep loess on the east bluff of the Mississippi River is highly uniform in texture throughout the entire length of that state. Nearly all of this soil was originally covered with deciduous hardwood forest. The analyses[1] of eight samples of this soil from the extreme south end of Illinois averaged 1.11 per cent of organic matter; four samples of the same stratum of the same soil

[1] MOSIER, J. G., and A. F. GUSTAFSON, "Soil Physics and Management," p. 145, J. B. Lippincott, Philadelphia, 1917.

type from the extreme north part of the state averaged 3.86 per cent of organic matter. Eighteen samples of timbered silt loam from the southern part of the state contained 1.5 per cent in the surface (0 to 6⅔ inches) and 0.58 per cent of organic matter in the subsurface (6⅔ to 20 inches). Comparable data for similar timber soils in the northern part of the state were 2.40 per cent in the surface and 0.96 per cent in the subsurface soil Samples of brown silt loam, a prairie soil, collected in central Illinois, contained 4.5 per cent of organic matter in the surface stratum, compared with 6.1 per cent in samples of the same stratum of the same soil type taken in the north part of the state. These data indicate that conditions for the accumulation of organic matter in the soil and for the control of erosion are more favorable in the North than in the South. And high altitude has the same general effect on the accumulation of organic matter as does high latitude.

Barnyard Manures.—All barnyard manure available should be conserved and spread on the land in moderate applications. On dairy farms 10 tons or loads to the acre once in three or four years gives best results for the manure ordinarily available. On grain or on mixed grain and livestock farms relatively less manure is produced than on dairy or other livestock farms. On such farms, therefore, the rate of application must be reduced to 8, or even to 6, tons to the acre, and it may be necessary to increase the interval between applications. If the manure supply is insufficient to provide a minimum of 6 or 8 tons to the acre once in four years, however, it will be necessary to make some other provision for adding organic matter to the soil on rolling lands.

Manure put on in light, relatively frequent applications gives better yields of feed crops and at the same time is more effective in controlling erosion than if used in heavy applications at long intervals. Eight tons of manure an acre every four years, for example, is more effective both as a fertilizer and in helping control soil losses than is 16 tons applied once in eight years.

Fall application of manure on slopes, particularly on heavy soils, results in some loss of plant nutrients from the manure. But such loss is offset to a large extent by the mechanical protection of the soil against erosion which the manure may afford on sloping lands.

Both from the standpoint of erosion control and from that of its use as fertilizer, uniform distribution of the manure is desirable.

And more uniform application is attained by means of a manure spreader than by hand.

Crop Residues.—The returning of crop residues[1] to the soil is one of the inexpensive ways of contributing to the maintenance of a good supply of fresh, active organic matter in the soil. The benefits to soils and crops resulting from the addition of organic matter are well known.[2] Residues add nitrogen as a nutrient for crops as well as fresh organic matter for the improvement of the tilth of the soil. Leaving such residues as small-grain stubble and corn and other stalks on the land over fall and winter protects the soil in a measure from the beating action of raindrops. Moreover, this type of organic materials aids in checking the flow of water on slopes and consequently in reducing the loss of soil, organic matter, and plant nutrients by erosion.

The greatest practicable return of all kinds of crop residues to the soil is advisable. Burning residues such as cotton, corn, or broomcorn stalks, and small-grain stubble is generally inadvisable. Sometimes residues are infected with serious plant diseases or infected with chinch bugs, European corn borers, or other insects. In that case the burning of residues may be warranted, but only, however, if burning actually aids materially in the control of such pests. Farmers frequently burn residues for the purpose of killing weed seeds. Spring burning of residues usually is not effective in killing weed seeds, however, because the seeds have shattered and fallen to the ground before the residues are dry enough to burn. Moreover, the seeds are moist and lie in close contact with the soil; consequently, such seeds are seldom injured because insufficient heat is developed. The usual result is the destruction of valuable organic matter, but the desired reduction in the number of weed seeds is seldom attained.

Green Manures.—Crops grown for the specific purpose of being plowed under in the green state for adding organic matter to the

[1] The term *crop residues* is used to designate such leftovers from crops as the stalks of corn, cotton, broom corn, and sorghums; the stubble and, in some areas, the straw of small grains and from the hulling of legume seeds; potato vines, stubble and aftermath of meadows, and the nonedible portion of vegetable crops.

[2] The beneficial effects of organic matter in soils are discussed by A. F. Gustafson, in Nitrogen and Organic Matter in the Soil, *Cornell Ext. Bull.* 201, revised ed., 1933.

soil are termed *green-manure crops* or green manures.[1] Green-manure crops can be grown under many soil and cropping conditions. Excellent use of green-manure crops can be made in the South. The growing season in that section is usually long enough for the production of the money crop such as cotton, tobacco, potatoes, vegetables, corn, or sweet potatoes and for

FIG. 50.—Tennessee No. 76 lespedeza. Lespedeza is a very valuable green-manure, feed, and erosion-control crop in the area to which it is adapted. It makes a good, protecting growth on unproductive soils. (*Courtesy of Tennessee Agr. Exp. Sta.*)

the growing of a green-manure crop for turning under during the same year.

It is desirable to use as a green-manure crop one or more legumes, preferably in a mixture with nonlegumes, wherever feasible. Perennial or biennial legumes are to be preferred to

[1] Additional information on green manuring is given by C. V. Piper and A. J. Pieters, in Green Manuring, *U. S. Dept. Agr., Farmers' Bull.* 1250, 1922; see also A. J. Pieters, "Green Manuring," John Wiley & Sons, Inc., New York, 1927.

the annual ones. Because the long-lived legumes produce more top growth and larger root systems, they add more organic matter to the soil than do the annuals. Moreover, thoroughly inoculated perennial and biennial legumes fix more atmospheric nitrogen than do the annual ones. And the addition of organic matter and nitrogen to many soils is sorely needed because the original meager supply of both organic matter and nitrogen has been severely depleted (Figs. 50 and 51).

FIG. 51.—Winter barley, after lespedeza on left, after corn on right. The barley on the right followed three years of corn, that on the left, three years of Korean lespedeza. The third year the lespedeza was plowed under in full growth. The barley following lespedeza controls erosion, but much washing takes place over winter in a growth such as that of the barley after corn. Previous cropping and fertilization were comparable. (*Courtesy of W. C. Etheridge, and C. A. Helm, Missouri Agr. Exp. Sta.*)

Cover Crops.—Crops grown mainly for covering the soil during late fall, winter, and early spring are designated *cover crops.* No definite distinguishing characteristic exists between cover and green-manure crops. The same plants are used for both purposes. By covering the soil these crops protect it from the detrimental effects of the beating action of raindrops and, by means of extensive root systems, hold the soil against washing. In addition, cover crops take up and hold plant nutrients against

[1] Additional information on green manuring is given by C. V. Piper and A. J. Pieters, in Green Manuring, *U. S. Dept. Agr., Farmers' Bull.* 1250, 1922; see also A. J. Pieters, "Green Manuring," John Wiley & Sons, Inc., New York, 1927.

loss from the soil. Nutrients that without the cover crop would have been lost from the soil are converted into fresh, active organic matter. This is added to and becomes a part of the soil when the land is plowed and prepared for the succeeding crop. The plant nutrients are then given up for use by crops (Figs. 52 and 53).

FIG. 52.—Sweet clover for green manure. Sweet clover, a valuable leguminous green-manure crop in the northern states. This growth is ready to turn under.

At the Pee Dee Experiment Station in South Carolina, Hall, Albert, and Watson seeded Austrian winter peas and vetches in the cotton middles early in September with a five-disk grain

TABLE 14.—YIELDS OF DRY MATTER AND NITROGEN IN TOP GROWTH OF VARIOUS COVER CROPS, 1929–1933*
(Five-year average)

Crop	Dry matter to the acre, pounds	Nitrogen to the acre, pounds
Austrian winter peas	1,860	56
Monantha vetch	2,733	91
Hairy vetch	2,129†	69
Hungarian vetch	1,787	52

* HALL, E. C, W. B. ALBERT, and S. J. WATSON, Winter Cover Crop Experiments at the Pee Dee Experiment Station, *South Carolina Agr. Exp. Sta., Circ.* 51, p. 12, 1933.

† Poor seed gave poor stand and low yield in 1932.

FIG. 53.—Peas and barley. Canada field peas and barley are a good mixture for a late spring green-manure crop over a wide north-south range.

FIG. 54.—Austrian winter peas. The Austrian winter pea has come into use rapidly in the South. This crop deserves a large place in southern agriculture. (*Courtesy of U. S. Soil Conservation Service, Oklahoma.*)

drill. In the later years reported, 40 pounds of the peas and 30 pounds of the vetches to the acre were used. The yield data were taken each year within a few days of Mar. 20, after which the cover crop was disked in, and the land plowed and prepared for cotton. The weight of dry matter and of nitrogen in each of the cover crops is given in Table 14 (see Fig. 54).

The effects of the cover crops and of additions of sodium nitrate on the yield of seed cotton are shown in Table 15.

TABLE 15.—AVERAGE YIELDS OF SEED COTTON TO THE ACRE FOLLOWING COVER CROPS, WITH AND WITHOUT SODIUM NITRATE, 1929–1932*
(Four-year Average)

Cover crop and treatment	For four years, pounds	Increase for cover crop, pounds	Increase for sodium nitrate, pounds
No cover crop	1,505		
Austrian winter peas	1,604	99	
Monantha vetch	1,735	230	
Hairy vetch	1,778	273	
Hungarian vetch	1,677	172	
No cover crop +200 lb. $NaNO_3$	1,654	...	149
Austrian winter peas +200 lb. $NaNo_3$	1,738	84	134
Monantha vetch +200 lb. $NaNO_3$	1,829	175	94
Hairy vetch +200 lb. $NaNo_3$	1,859	205	81
Hungarian vetch +200 lb. $NaNO_3$	1,790	136	113

* HALL, E. C., W. B. ALBERT, and S. J. WATSON, Winter Cover Crop Experiments at the Pee Dee Experiment Station, *South Carolina Agr. Exp. Sta., Circ.* 51, p. 13, 1933.

The fertilizer applied for cotton each year was reduced progressively in nitrogen and potash from 750 pounds of 4-8-4 to the acre, used in the years 1924–1929, to 540 pounds of 4-10-4 in 1931 and 1932. Attention is called to the marked increases in yield of seed cotton produced by the hairy and monantha vetches and to the relatively minor increases produced by the application of 200 pounds of sodium nitrate a year to the acre in addition to turning under the cover crops.

The effects of various cover crops on cotton and corn yields at the Coastal Plain Experiment Station at Tifton, Ga., as reported by Alexander, are shown in Table 16.

The very pronounced effect of the return of organic matter as compared with "none" is more notable with cotton. Austrian peas without nitrogen in the fertilizer nearly doubled the yield of seed cotton produced on the plot without the peas. The vetches and the rye with nitrogen produced very large increases in yields of cotton. Without the addition of fertilizer nitrogen, however, rye reduced the yield of corn below that on the check

TABLE 16.—YIELDS OF COTTON AND CORN FOLLOWING SOIL-IMPROVEMENT CROPS IN GEORGIA*
(Seven-year average)

Crops	Seed cotton per acre, pounds		Corn per acre, bushels	
	9-0-5 fertilizer	9-3-5 fertilizer	10-0-4 fertilizer	10-2-4 fertilizer
Austrian winter peas	1,233	1,199	57.1	52.7
Monantha vetch	1,064	1,305	48.0	50.3
Hairy vetch	961	1,215	46.9	50.4
Rye	956	1,167	34.1	38.4
None	690	913	37.9	39.7

* ALEXANDER, E. D., Austrian Winter Peas and the Vetches, *Georgia Agr. Ext., Bull.* 453, p. 4, 1935.

plot. This loss results, no doubt, from the withdrawal of nitrogen temporarily from the soil by the bacteria which brought about the decomposition of the rye in the soil. The crop is thus deprived of the use of this nitrogen in a critical stage of growth. Turning the rye under at an earlier stage when it contains a higher proportion of nitrogen may be expected to remedy this difficulty.

This experimental work and that in South Carolina reported in Tables 14 and 15 was done on the Coastal Plain, but the general results obtained are believed to be equally applicable to the Piedmont. Such additions of organic matter to the soil may be expected not only to increase crop yields but also to aid in the control of soil erosion in the easily eroded Piedmont.

Wolfe and Kipps report a decade's work in growing both summer and winter cover crops in Virginia for the production of organic matter and for the fixation of nitrogen for the use of corn and wheat. Crimson clover and vetch, soybeans, and buckwheat all increased the yield more when these were turned under than when they were taken off for hay. Rye, however, decreased the yield slightly when turned under as it did in the Georgia work (Table 16). These data (Tables 17 and 18), like those from Georgia and South Carolina, emphasize the benefits obtained by nonlegume crops from the plowing down of farm-produced organic matter (Figs. 55, 56, and 57).

FIG. 55.—Rye and vetch at a good stage for plowing under. Not only do rye and vetch supply organic matter, but the vetch fixes atmospheric nitrogen. The mixture controls erosion through the winter and early spring.

TABLE 17.—THE EFFECTS OF TURNING UNDER CRIMSON CLOVER AND VETCH AND RYE ON THE AVERAGE YIELD PER ACRE AND PERCENTAGE OF MARKETABLE GRAIN*
(Eight-year average)

Treatment	Yield, bushels	Marketable corn, per cent
Crimson clover and vetch cut for hay	41.19	84.30
Crimson clover and vetch turned under	45.90	87.53
Check—no cover crop	30.74	78.06
Rye turned under	24.07	67.24
Rye cut off	26.07	73.37

* WOLFE, T. K., and M. S. KIPPS, The Effect of Rotation, Fertilizers, Lime and Organic Matter on the Production of Corn, Wheat, and Hay, *Virginia Agr. Exp. Sta., Bull.* 253, p. 45, 1927.

FIG. 56.—Wheat at a good green-manure stage. Like the rye and vetch, wheat controls erosion throughout the fall, winter, and early spring to excellent advantage.

FIG. 57.—Soybeans in Illinois. Being a legume the soybean is a good green-manure crop in its area. This photograph was taken in 1913.

TABLE 18.—THE EFFECT OF SOYBEANS AND BUCKWHEAT ON THE YIELD OF WHEAT*

(Five-year average for wheat and six-year average for hay)

Method of handling	Yield per acre	
	Wheat, bushels	Hay, tons
Soybeans cut for hay	24.12	2.15
Soybeans turned under	30.80	
Check—no cover crop	19.84	
Buckwheat cut for hay	15.42	.72
Buckwheat turned under	23.88	

* WOLFE, T. K., and M. S. KIPPS, The Effect of Rotations, Fertilizers, Lime, and Organic Matter on the Production of Corn, Wheat, and Hay, *Virginia Agr. Exp. Sta., Bull.* 253, p. 49, 1927.

Maximum use of cover and green-manure crops is desirable in orchards, citrus and nut groves, vineyards, between the rows of bush fruits, on vegetable lands, and following potatoes, corn, cotton, and other intertilled crops in all sections that have soil, crop, and climatic conditions warranting the labor and expense

FIG. 58.—Crotolaria in Florida. This is the *Crotolaria striata*, a highly useful green-manure crop in the deep South. (*Courtesy of W. E. Stokes, Florida Agr. Exp. Sta.*)

of seeding them. Throughout the South, with its relatively long period of temperatures too low for active growth of crops but sufficiently high for slow decay of organic matter, green-manure and cover crops are highly useful. The specific use of cover and green-manure crops and their place in representative rotations are given under Rotations (see Fig. 58).

Important Winter Legumes for the Cotton Belt.—Following are some important winter legumes for the cotton-growing area of the United States.[1]

Name	Section
Winter green-manure crops	
Legumes	
Hairy vetch (*Vicia villosa*)	General
Austrian winter pea (*Pisum arvense*)	General
Smooth vetch (*Vicia villosa* var.)	General
Woolly-pod vetch (*Vicia dasycarpa*)	General
Crimson clover (*Trifolium incarnatum*)	Northern part
Monantha vetch (*Vicia monantha*)	Southern part
Tangier pea (*Lathyrus tingitanus*)	Southern part
Southern bur clover (*Medicago arabica*)	Southern part
Tifton bur clover (*Medicago regidula*)	Southern part
Summer green-manure crops	
Legumes[2]	
Alfalfa (*Medicago sativa*)	In western orchards
Common sesbania (*Sesbania macrocarpa*)	Southwest
Cowpeas (*Vigna sinensis*)	South and southern Midwest
Soybeans (*Soja max*)	Wide range
Red clover (*Trifolium pratense*)	Wide range—North
Sweet clover (*Melilotus alba* and *M. officinalis*)	Wide range
Lespedeza:	
Kobe and Tennessee 76 (*Lespedeza striata*)	Wide range—South to southern Midwest
Korean lespedeza (*Lespedeza stipulacea*)	Wide range—South to southern Midwest
Crotolaria (*Crotolaria spectabilis*)	Lower South

[1] Additional information is available in Winter Legumes for Green Manure in the Cotton Belt, by Roland McKee and A. D. McNair, *U. S. Dept. Agr., Farmers' Bull.* 1663, 1936.

[2] Additional information is available in Summer Crops for Green Manure and Soil Improvement, by Roland McKee, *U. S. Dept. Agr., Farmers' Bull.* 1750, 1935.

Crotolaria (*Crotolaria striata*)	Lower South
Florida beggarweed (*Meibomia purpurea*)	Lower South
Deering velvet beans (*Stizolobium deeringanium*)	South
Nonlegumes	
Sudan grass (*Sorghum vulgare* var. *sudanensie*)	Wide range
Buckwheat (*Fagopyrum esculentum*)	North
Spring grains also are useful	Wide range

FERTILIZATION FOR HIGH YIELDS AND ADEQUATE SOIL COVER

High crop yields are essential for economic success in agriculture, particularly under conditions of high fixed costs, including taxes and interest charges or rent. On well-fertilized productive soils, crops make relatively rapid growth which soon affords protection to the soil. Luxuriant crops produce large root systems which hold the soil against erosion. And the top growth protects the soil in a sense the same as do the leaves of the forest trees.

Moreover, a large crop leaves relatively more residues for the protection of the soil after the crop is harvested than does a small one grown on poor, unfertilized land. Furthermore, the larger crop adds much more organic material as roots, which are thoroughly distributed throughout the soil, and more aboveground residues to be turned under than does the small crop. Fertilization, particularly the maintenance of a high content of available phosphorus by means of its effects on crops, aids generally in keeping the soil in a high state of productivity, a condition favorable to the conservation of the soil against erosion.

FOLLOWING A DEFINITE ROTATION OF CROPS

The general benefit of practicing a definite rotation, or regular sequence of crops, has already been discussed briefly. Additional benefits from the standpoint of erosion control follow the growing of crops in regular rotation.

Lewis and Riesbol report the results of five years of comparison between cotton that was grown continuously and a three-year rotation of cotton, wheat, and sweet clover at Guthrie, Okla. The average rainfall for these five years was 33 inches. The data are given in Table 19.

TABLE 19.—EFFECT OF THREE-YEAR ROTATION COMPARED WITH CONTINUOUS COTTON, GUTHRIE, OKLA., 1930–1934*

Plot No.	Crop treatment	Water-loss runoff, per cent	Soil loss per acre, tons
3	Cotton—continuous	15.33	25.47
5	Cotton in rotation	13.83	15.37
6	Wheat in rotation	14.52	1.72
4	Sweet clover in rotation	8.31	0.62
	Average for rotation	12.22	5.90

* LEWIS, H. G., and H. S. RIESBOL, Outline of Investigations and Summary of Results: 1930–1935. *Red Plains Soil Erosion Exp. Sta., Guthrie, Okla., U. S. Dept. Agr.*, SCS-AP-6, p. 4, 1936.

Two points stand out. On an average, the loss of soil was only three-fifths as great from cotton land in a three-year rotation as from continuous cotton. Losses of soil from the wheat and sweet-clover plots were so slight as to reduce the average annual loss from the rotation plots to but one-fifth of that from the continuous-cotton plot. Water losses from the rotation plots were somewhat greater than might have been expected in view of the marked reduction in the average loss of soil. Practicing this three-year rotation, however, reduced the loss of water by one-fifth on the average.

Miller and Krusekopf report the results of controlling erosion on actively cultivated lands in Missouri. This is one of the earliest and most excellent pieces of continuous research on erosion control in the United States. The data showing the effects of rotation as compared with continuous cropping are given in Table 20.

As an average for the fourteen years, continuous corn lost 19.7 tons of soil, and continuous wheat 10.1 tons from an acre a year. In comparison, continuous bluegrass (*Poa pratensis*) lost 0.3 ton a year; and a three-year rotation of corn, wheat, and clover lost 2.7 tons of soil a year on the average for the rotation.

During its growing season the continuous-corn plot lost 2.5 times as much water and 4.7 times as much soil as did the three-year rotation plot. The continuous-wheat plot lost 3.74 times as much soil as did the rotation plot. These data show the highly desirable effect of this rotation on the conservation of

TABLE 20.—RELATION OF CROPPING SYSTEM TO SOIL AND WATER LOSSES IN MISSOURI*
(Fourteen years' results, 1918–1931)

Cultural and cropping systems	Average yearly loss		No. of years to erode 7 inches of soil
	Rainfall, per cent	Soil, tons	
Fallow, plowed 8 inches	30.3	41.0	24
Continuous corn	29.4	19.7	50
Continuous wheat	23.3	10.1	100
Corn, wheat, clover	13.8	2.8	368
Continuous bluegrass	12.0	0.3	3,043

* MILLER, M. F., and H. H. KRUSEKOPF, The Influence of Systems of Cropping and Methods of Culture on Surface Runoff and Soil Erosion, *Missouri Agr. Exp. Sta., Res. Bull.* 177, p. 22, 1932.

water and soil as compared with that of continuous corn or wheat. Continuous bluegrass, on the other hand, retains the soil better than does the rotation. Attention is called to the third column of Table 20 showing the number of years required for the top 7 inches of soil to wash away under these cultural and cropping systems in Missouri.

In Table 21 Wolfe and Kipps report a materially higher yield of corn in a three- than in a four- or five-year rotation. Wheat and hay yields, on the other hand, are not materially changed.

TABLE 21.—THE AVERAGE YIELDS PER ACRE OF THE VARIOUS CROPS IN THREE-, FOUR-, AND FIVE-YEAR ROTATIONS*

Crop	Rotation		
	three-year	four-year	five-year
Corn, bushels	39.63	49.98	49.79
Soy beans, tons	.....	.....	2.51
Wheat, bushels	24.62	22.45	24.34
Clover, tons	2.30		
Clover and grass (first year), tons	.....	1.95	1.81
Grass and clover (second year), tons	.....	1.70	1.78

* WOLFE, T. K., and M. S. KIPPS, The Effects of Rotations, Fertilizers, Lime, and Organic Matter on the Production of Corn, Wheat, and Hay, *Virginia Agr. Exp. Sta., Bull.* 253, p. 13, 1927.

Owing to the introduction of soybeans into the five-year rotation, it is not strictly comparable with the four-year one.

A Potato-feed-crop Rotation for Parts of the Northeast States. A simple four-year rotation which consists of potatoes the first year, spring grain the second, clover the third, and mixed clover and grasses—largely timothy—the fourth year is being followed in potato sections of southern New York and under similar conditions in other northeast states. Although admittedly not the ideal rotation for the production of potatoes owing to the insect enemies of that crop harbored by the grass sod, this rotation is, and will continue to be, followed on rolling to steeply sloping lands until a rotation better from every standpoint is developed.

The sod is all too generally plowed in the fall, and some erosion often takes place before potato-planting time the following spring. Under many conditions, early spring plowing of the sod is preferable because then only a short period for possible erosion elapses between plowing and planting time for potatoes. In fall-plowed soil the grass roots are killed in the fall and are partly decomposed before potato-planting time. In the spring-plowed sod, on the other hand, the grass roots are killed only a short time before the potato crop is planted. The grass roots, therefore, are still in a condition to and do hold the soil against washing during the early part of the potato-growing season.

Legume roots are less effective in holding the soil than are those of nonlegumes, because the former decay more rapidly than do the latter. The roots of such grasses as timothy, redtop, orchard grass, and Bermuda grass are relatively persistent in the soil and aid materially in checking erosion. Legumes, however, are essential in all good meadow-seeding mixtures.

In this potato rotation the soil is plowed only once in four years and is then protected to a marked degree by the grass roots, as already stated. The critical period for erosion is during late fall, winter, and early spring following the harvesting of the potato crop. The potato digger leaves the soil loose and smooth, since the ridges made during potato cultivation are leveled by the potato harvester. In this loose, level condition the soil is particularly susceptible to erosion.

In the latitude of southern New York at elevations of more than 1,200 feet the harvesting of late potatoes is seldom completed until immediately before the soil freezes. In fact, the soil freezes before digging is finished in some years. In this area the use of rye promises to aid materially in erosion control on hilly land following late potatoes.

Two methods of seeding rye appear feasible. With either one the steepest potato land is harvested first, the less erosive land later, and the "levelest" or least erosive soil is harvested last. From the standpoint of erosion control the early strains of late potatoes are most desirable.

1. Rye is seeded each evening on the area from which the potatoes have been dug during the day. A grain drill of the disk type is more desirable than one of the hoe type because the latter tends to rake the potato vines together, whereas the disk drill runs over them. In fact, the disk drill presses the vines into the soil, and thus the vines aid in erosion control. Potato vines, in common with other crop residues, are needed in the soil for their organic matter and plant nutrients and consequently ought not to be burned.

The purpose of following the digger so closely with seeding of the rye is to give the rye the benefit of all of the growing weather that may follow in order to obtain the maximum fall growth. The rye is drilled as nearly as possible on contour lines in order to avoid bringing about any concentration of water which might lead to gully formation along the drill rows.

The grain drill makes small furrows 7 or 8 inches apart, whereas the digger leaves the soil relatively smooth. The drill furrows hold water from rain or snow and cause it to percolate into unfrozen soil, but from the soil smoothed by the digger, once the soil has settled, the water flows unobstructed down the slope. Leaving the soil relatively rough aids in checking the runoff of water and in the control of erosion.

2. The other method is to seed the rye broadcast several weeks before digging the potatoes. Rain may be expected to cover the seed and bring about early germination of the rye. For the rye to be from 2 to 3 inches in height when the potatoes are harvested is desirable. The potato digger will cover and kill some of the rye, but a strip of approximately a foot in width may be expected to survive and grow. For rye seeded in this way to be successful the potato ridges should be very low lest too much of the rye be covered too deeply. A field treated in this manner

will have strips of rye as a cover crop alternated with bare soil. Any soil washed from the bare strip during the fall, winter, and early spring is caught and held by the rye.

The possible advantage of seeding rye in this way in comparison with seeding with the drill after potatoes have been harvested lies in the greater probability of obtaining sufficient growth of the rye to warrant its use. This second method is used only where little growing weather can be expected after potatoes have been harvested. Seeding rye with a grain drill after potatoes have been dug is the better method wherever the length of the growing season warrants the expectation of a growth of 2 inches before the soil freezes.

Hairy or winter vetch (*Vicia villosa*) can be added to the rye with highly beneficial results in many sections. In common with other legumes, vetch seed requires inoculation.

The following spring, the land that produced potatoes the preceding year is stirred with the spring-tooth or the disk harrow in preparation for the seeding of the spring-grain crop. Some rye may survive this tillage, but it will do little harm in the spring grain. The clover-grass meadow-seed mixture is applied in a separate operation and covered with the weeder. In parts of the northeastern states the spring grain comes up and establishes itself quickly and holds the soil against erosion. During the clover and the grass years of the rotation, essentially no erosion takes place.

Corn may be grown the first year in this four-year rotation in place of potatoes. As has been pointed out, corn is a less erosive crop than potatoes. Consequently, somewhat less precaution is required for controlling erosion in growing corn than in the production of potatoes.

A Feed-crop Rotation for the Northeast States.—A suitable feed-crop rotation for the northeast states on moderately rolling land is one of four years: corn or other intertilled crop, the first year; spring grain, oats, barley, or mixed oats and barley, the second; clover and timothy, the third; and timothy and clover, the fourth.[1] Under average soil conditions in the Northeast 6 pounds of red clover, 4 of alsike, both of which need inocula-

[1] Feed-crop rotations suitable for various parts of New York and applicable throughout the northeast states are found in Part I of *Cornell Univ. Agr. Exp. Sta., Bulls.* 538, 567, 570, 600, 630, and 639 by A. F. Gustafson.

tion, and 8 pounds of timothy to the acre is a commonly recommended rate of seeding.

On steeper and relatively more erosive land a longer rotation is conducive to better erosion control. On soils adapted for the growth of alfalfa which contain sufficient lime, or to which adequate lime has been applied, 6 pounds of inoculated hardy alfalfa seed to the acre is used in place of the alsike clover in the foregoing meadow-seed mixture. The alfalfa-timothy mixture may be kept productive over a fairly long period of years if the soil was adequately limed before seeding the alfalfa and timothy, and if sufficient phosphorus is added from time to time. The quantity of phosphorus needed varies in a measure with previous fertilization. The addition of phosphorus at the average annual rate of 30 pounds of phosphoric acid (approximately 200 pounds of 16-per-cent superphosphate or 100 pounds of 32-per-cent superphosphate) to the acre materially improves the yield of alfalfa. The application of 400 pounds of 16-per-cent superphosphate or its equivalent in other analyses every two years, or 600 pounds every three years, is perfectly feasible and gives good results.

If the so-called insoluble finely ground rock phosphate is used, larger quantities are required both as initial and as subsequent applications, 1,000 to 2,000 pounds of rock phosphate to the acre often being used.

In these rotations the clover, as well as the alfalfa-timothy mixtures, requires adequate additions of phosphorus. Stress is laid upon the need of phosphorus by legumes here because the legume is the key to erosion control in these rotations. Inoculated legumes fix nitrogen which is used by the grasses. The nitrogen helps to keep the grasses thrifty; and the using up, or the draining off, as it were, of the nitrogen fixed by the legumes on the part of the grasses keeps the legume working through the nitrogen-fixing organisms in the nodules on its roots. Strong evidence exists to the effect that long-lived legumes, alfalfa in particular, live and continue in a productive condition longer when growing in association with nonlegumes, because the latter do use up and take away the nitrogen fixed by the alfalfa. A mixture of legumes and grasses, all of which are adapted to the soil and the climate in which they are grown, is ideal for complete erosion control during the years that the land is in meadow.

Corn Rotation for the Midwest.—A commonly used four-year rotation for intensive corn production in the Midwest is: corn the first and second years, oats the third, and clover the fourth year. On the more rolling and more erosive lands, or on soils of low organic content such as timbered areas, a three-year rotation will afford superior erosion control. The latter rotation has but one year of corn and is then one of corn, oats, clover. Planting soybeans on part or all of the second-year corn acreage aids in the maintenance of high yields and may help slightly in reducing erosion.

A still further and a very desirable variation in the four-year rotation for thin rolling lands is: corn the first year, soybeans the second, wheat the third, and clover the fourth year. The soil is occupied by a legume during all of the fourth and during a part each of both the second and third years of this rotation. This is a desirable situation in that the legumes help in keeping up the supply of fresh organic matter and nitrogen and as a result aid in maintaining the productivity of the land. High yields, as previously stated, by means of the production of large root systems and of luxurious top growth protect the soil against erosion.

For the more erosive corn lands, the length of the rotation may be extended by means of one or more years of timothy and other grasses following the clover. Or alfalfa and timothy may displace the clover in these rotations and be cut additional years as may be desired or as may better satisfy the feed requirements of the farm.

Cotton Rotations for the South.—Any one cotton rotation will not be wholly ideal for the entire cotton-growing area of the United States, but the general principles of cotton rotations can be presented in a few specific examples.

Alexander suggests the following rotations for Georgia:[1]

Cotton-forage-corn rotation, three years.

FIRST YEAR

Spring, seedbed preparation for cotton
Summer, cotton
Fall, winter legume seeded with small grain

[1] ALEXANDER, E. D., Austrian Winter Peas and the Vetches, *Georgia Agr. Exp. Sta., Bull.* 453, pp. 19–24, 1935.

SECOND YEAR

Spring, winter legume and small grain cut for hay or grazed
Summer, summer legume for forage
Fall, winter legume

THIRD YEAR

Spring, winter legume turned under
Summer, corn interplanted with summer legume
Fall, land turned and laid off on contour for cotton

The foregoing may be changed to a *cotton-small grain-corn* rotation of three years by omitting the winter legume in the fall of the cotton year, that is, by seeding oats or other small grain alone and by harvesting small grain for grain the second year instead of for forage, as indicated in the first rotation.

Cotton-corn rotation, two years.

FIRST YEAR

Spring, seedbed preparation for cotton
Summer, cotton
Fall, winter legume

SECOND YEAR

Spring, winter legume turned under
Summer, corn interplanted with legume
Fall, land turned for cotton as in third year of first rotation

This rotation is more intensive than either of the others outlined.

The outstanding features of these rotations are the frequent recurrence of legumes and the very brief period of the year during which the soil is bare or not protected from erosion by a crop. In the eleven additional rotations suggested by Alexander, legumes occupy an equally important place. Such frequent use of legumes in Georgia and in neighboring states as well is highly advantageous to agriculture, both by increasing crop yields and by reducing soil losses through erosion.

Cotton Rotation for South Carolina.—A two-year *cotton-corn* rotation for South Carolina is suggested by Hamilton.[1]

[1] HAMILTON, R. W., Soil Building Rotation Demonstration, *Clemson Agr. Coll., Agr. Exp. Sta., Circ.* 99, p. 4, 1929.

FIRST YEAR

Cotton, Austrian winter field peas, Abruzzi rye or hairy vetch (vetch preferred)

SECOND YEAR

Corn, with Atootan soybeans in drill and Austrian field peas or rye.

In South Carolina, as in Georgia, a most commendable wide use is made of legumes in these rotations. High crop yields and relatively good control of erosion should result.

A Commonly Used Virginia Rotation.—In Virginia a commonly used five-year rotation is corn, with crimson clover and vetch, seeded at the time of the last cultivation, the first year; soybeans, seeded after the clover and vetch have been plowed under, the second; wheat the third; clover and grass the fourth; and grass and clover for hay the fifth year. Following this rotation gives increasing crop yields and a gain in the nitrogen and in the active organic-matter content of the soil. Furthermore, by keeping the soil covered by a crop nearly all of the time, this rotation is decidedly helpful in the control of erosion.[1]

The yields obtained in the five-year rotation are given in the last column of Table 21. The four-year rotation was similar to the five-year except that the soybeans were omitted from the former. And the three-year was similar to the four-year rotation except that clover constituted the third year of the rotation, which was corn, wheat, clover, with the cover crop of crimson clover and vetch as in the other rotations. The corn yield in the three-year rotation was inferior to that in the four- and five-year ones.[2]

Tobacco-corn Rotations for Kentucky.—Some excellent rotations for Kentucky are recommended by Roberts.[3] A rotation of six years or two rotations of three years each are presented here: (1) tobacco the first year, wheat the second, and clover the third year; (2) corn the first year, wheat the second, and clover and grass the third year.

[1] See footnote to Table 21.

[2] Additional information bearing on details of tobacco, cotton, and peanut rotations in Virginia is found in A Handbook of Agronomy, *Virginia Polytec. Inst., Bull.* 97, 1934.

[3] ROBERTS, GEORGE, Soil Management for Kentucky, *Kentucky Ext. Circ.* 272, pp. 47–48, 1934.

The tobacco, being the cash crop, may be grown on such acreage as is desired, and the feed crops required grown on other fields on the farm. The meadow seeding in the wheat of the corn rotation may be adjusted between legumes and grass to suit the needs of the tobacco that follows them when the rotations are alternated. In these rotations the soil is exposed to erosion only a small part of the time. Very good control of erosion, therefore, may be expected under the cropping conditions of these rotations.

Lespedeza-grain Rotations in Missouri.—Etheridge and Helm recommend rotations of Korean lespedeza with grain: (1) oats seeded with lespedeza, (2) wheat, (3) winter barley, and (4) rye seeded with Korean lespedeza. Each crop may be grown continually on a given field or such of these grains as it is desired to grow may be rotated.[1]

A lespedeza-corn rotation may be one of three years: corn, seeded with Korean lespedeza at the time of the last cultivation, the first year; oats, lespedeza the second year; and wheat with lespedeza the third year. After a few years of growing Korean lespedeza in this way it reseeds itself under favorable conditions sufficiently to give a good stand. This rotation may be extended as desired by pasturing the lespedeza or by making hay of it. These rotations bring about a high degree of erosion control because the soil is protected by lespedeza throughout the late fall, winter, and early spring every year.

Lespedeza rotations are not displacing the conventional corn-oats-clover rotations entirely. Lespedeza sown in the corn of the conventional rotation materially improves the control of erosion on sloping lands in the sections to which lespedeza is adapted.

A Sorghum-wheat Rotation for Central Kansas.—A practical four-year rotation for central Kansas suggested by Laude and Swanson[2] might include sorghum the first year, barley or oats the second, and wheat both the third and the fourth years. They report the average yield for seven years at the Fort Hays substation as 30.6 bushels to the acre for kafir, 27.7 for barley. 27.1 for first-year wheat, and for second-year, wheat, 24.9 bushels,

[1] ETHERIDGE, W. C., and C. A. HELM, Korean Lespedeza in Rotations of Crops and Pastures, *Missouri Agr. Exp. Sta., Bull.* 360, pp. 6–12, 1936.

[2] LAUDE, H. H., and A. F. SWANSON, Sorghum Production in Kansas, *Kansas Agr. Exp. Sta., Bull.* 265, pp. 39–40, 1933.

In northwestern Kansas more barley and less wheat, and in southwestern Kansas more sorghum and less wheat with fallowing at regular intervals, is recommended.

Corn-wheat Rotations for Eastern Kansas.—For eastern Kansas, Salmon and Throckmorton[1] state that a popular rotation consists of corn the first and second years, oats the third, sweet clover the fourth, and wheat the fifth year. Under farm conditions requiring red instead of sweet clover the rotation may be changed slightly to corn, oats, wheat, red clover. On especially good wheat soils, one year of the corn may be displaced by an additional year of wheat.

On the poorer or the more erosive soils of eastern Kansas and adjacent areas a rotation of corn, seeded to lespedeza at the final cultivation, the first year; oats the second; sweet clover the third; wheat the fourth; and red clover the fifth year may be effective in the control of erosion and at the same time produce wheat for sale and an abundance of feed for the farm livestock.

Further examples of rotations suited to the soils, crops, climate and feed-, and cash-crop requirements might be discussed, but the foregoing are sufficient to indicate the relations between crop rotations and soil-erosion control. Good soil management and proper crop-rotation practices are invaluable aids in the control of soil erosion.

[1] SALMON, S. C., and R. I. THROCKMORTON, Wheat Production in Kansas, *Kansas Agr. Exp. Sta., Bull.* 248, pp. 11–17, 1929.

CHAPTER IX

CONTOUR TILLAGE AND PLANTING ESSENTIAL

Next to protecting the soil by means of a thick cover of vegetation, the growing of crops on the contour or on level lines across the slope is one important way of controlling the loss of water and soil on moderate to steeply sloping lands. Contour cropping is inexpensive and can readily be practiced by most farmers. Much space is devoted to the details of contour farming on the following pages, including some repetition with specific examples, admittedly for the purpose of emphasis.

The extent to which contour farming can be depended on for the control of erosion varies greatly with the erodibility of the soil, the kind of crops grown, and the type of rainfall. Differences in the erodibility of soils has been discussed (pages 66 to 72). On lake-laid soils, such as the Dunkirk silty clay loam, for example, contour cropping with suitable rotation may afford a fair degree of erosion control under average normal rainfall in the lake region of New York on slopes up to, but certainly not much greater than, 3 per cent. On the other hand, erosion may be controlled successfully by contour rotation cropping on absorptive soils that do not erode easily on slopes up to 5 or 6 per cent. If close-growing crops predominate, and if little of the rainfall is of the heavy-shower type, fair control may be accomplished on somewhat steeper slopes.

CONTOUR TILLAGE AND SEEDING

Plowing is best done, as nearly as feasible, on the contour. In contour plowing, the little depressions which usually develop upon settling of the furrow slice are on the level. In that position these depressions hold rain water until it soaks into the newly plowed soil. If plowing is done up and down the slope, gullies result when sufficient water collects in such depressions. Any sloping depression across the up-and-down furrows concentrates sufficient water in them to cause severe gullying during heavy rains.

The side-hill, or two-way, plow is preferable to the ordinary plow on hilly land because the formation of dead furrows can be avoided by using it. Dead furrows left by the ordinary one-way plow have started the formation of many large gullies in various sections of the country. This is well illustrated in Fig. 4.

The operations of seedbed preparation should, in the main, follow contour lines. At times, however, cross harrowing is desirable in order to produce a level seedbed, particularly if the soil was rough following plowing. Up-and-down hill tillage with the spring-tooth harrow produces little ridges and furrows. During heavy rains, water collects in and flows down these depressions, and washing of the *rill*, or *shoestring*, type follows (Fig. 6). The same little furrows left by the spring-tooth or even by the peg-tooth harrow across the slope hold rain water and cause it to percolate into the soil.

Intertilled crops, such as cotton, corn, potatoes, beans, sorghum, and vegetables, should be planted on the contour in order that they may be cultivated on the level. Up-and-down hill cultivation leads inevitably to soil washing, often resulting in severe damage during quick heavy showers. Contour cultivation, on the other hand, brings about ponding of water between the rows. Heavy showers are often held without loss of water by these little terraces formed by cultivation. And because the water is held until it is taken up by the soil, no erosion occurs. Furthermore, the water which may be retained by contour tillage is often needed by the crop. If the water is held, the yield of crops usually is correspondingly increased. It has been noted that crop growth and resulting yields are higher on the middle and lower parts than on the upper part of slopes. The retention of more water on the lower than on the upper part of slopes is undoubtedly one factor in superior yields on the lower part of the slope (Fig. 59).

Another point worthy of note in passing is that a field, on which contour cultivation holds the water of rains, does not contribute water for flooding the valleys below.

If corn or other crops must be planted up- and downhill on irregular or compound slopes, the use of the peg-tooth harrow is advisable for filling the planter tracks. This aids in the avoid-

Fig. 59.—Near-contour rows on compound slopes. An effort was made to run these rows on the contour, but owing to the compound slopes, there is considerable fall in the rows. At these points the water breaks over, as in the foreground, and gully formation follows.

Fig. 60.—Corn rows up and down a slope. Much of the corn was washed out by the water concentrated in the planter tracks. Even the harrow-teeth marks led to gully formation. (*Courtesy of L. D. Baver, Missouri Agr. Exp. Sta.*)

ance of washing down the corn or other crop rows. If the planter tracks are not eliminated, heavy showers falling before the first cultivation sometimes wash corn out completely (Figs. 60 and 61). One of the writer's recollections of his early farm work is that of replanting corn with a hoe in planter-track gullies on a timber-soil hill farm in western Illinois. In that section, and in fact wherever the planting-season rain falls in heavy, dashing showers, the planter may well be followed by the peg-tooth

Fig. 61.—Final result of up-and-down hill corn planting. This photograph, taken in 1911, shows a field only a few years after the latest crop of corn was grown on it. The stalks still stood on some of the ridges.

harrow within a few days to eliminate the planter tracks. A source of loss is thus removed, because the yield of corn, and other crops as well, is reduced by this type of erosion even if the crop is replanted with the utmost care.

In common with other crops small grains are seeded to best advantage across the slopes. The little ridges and furrows made by the grain drill are thus on the contour. These furrows being only 7 or 8 inches apart hold much water until it percolates into the soil. Up-and-down hill seeding of grain, on the other hand, facilitates erosion down the drill furrows.

In time, rains level these useful drill ridges. Even after leveling occurs, however, the contour rows of grain plants check the flow of water down the slope so that much of it soaks into the soil and is held for future use instead of being lost as runoff. Up-and-down hill rows in comparison with contour rows of grain plants are far less efficient in holding water, because the former do not offer comparable resistance to the flow of water.

Among newly planted crops small grains seeded on the contour are particularly effective in retarding runoff and in holding water in the soil for use later in the season. The fact that absorption of a high proportion of the water falling in a heavy shower can be accomplished on sloping land in crop is of great interest to the general public owing to its important bearing on flood prevention.

Furthermore, meadow seedings are usually made as the grain is being drilled or shortly afterward. The grass seed is blown or washed into the drill furrows, and consequently the meadow plants are arranged in somewhat indefinite rows on contour lines. During the entire life of the meadow the contour grass rows definitely slow up the flow of water over the slope and bring about the absorption of a large proportion even of heavy showers. Up-and-down hill rows of grass are far less effective than are those on the contour in the conservation of rain water.

Contour plowing, seedbed preparation, seeding, cultivation, and harvesting require less power if done on the contour than if done up- and downhill. In fact, farmers recognize that contour, in contrast with up-and-down hill cropping is markedly easier on both team and driver. Once these benefits and advantages of contour cropping are fully understood and appreciated by farmers, they adopt the contour methods readily in many areas.

The rectangular fields so common in the Midwest and wherever the land was surveyed into square townships and these into square sections, which in turn were subdivided into square 40-acre fields, are a definite obstacle to the adoption of contour farming methods. It has seemed to be easier and more economical of time and power to plow, plant, cultivate, and harvest crops on lines parallel to or at right angles to the field boundaries than otherwise. The habit of farming parallel to roads and field boundaries has become firmly established in many sections.

As already pointed out, soils of uniform texture such as wind- or lake-laid ones wash or erode very easily. Slopes even so gentle as from 2 to 5 per cent suffer the loss of surface soil over a period of years. On slopes of 10 to 15 per cent on such soils, severe surface erosion and finally gullying takes place during a period of up-and-down hill farming. Much evidence has already accumulated to the effect that slopes of 5 to 10 per cent have lost so much surface soil that immediate steps should be taken to check such losses. Time and patience on the part of the personnel of all the agricultural education agencies will be required to bring about such changes in farming methods as will truly conserve the soil resources of the sloping and hilly lands of the country.

Benefits of Contour Planting.—Lewis and Riesbol report three-year average losses of water and soil from plots on which cotton was grown continuously. The average annual rainfall for the period was 34.7 inches. On one plot the cotton was planted up and down the slope, and across the slope on the other. The data are given in Table 22.

TABLE 22.—EFFECT OF DIRECTION OF ROWS ON WATER AND SOIL LOSSES (Continuous cotton, three-year average, 1932–1935)*

Direction of rows	Average slope, per cent	Runoff, per cent	Soil loss per acre, tons
Up-and-down slope	6.75	13.27†	56.64
On contour	6.87	10.94†	24.25

* LEWIS, H. G., and H. S. RIESBOL, Outline of Investigations and Summary of Results: 1930–1935, Red Plains Soil Erosion Exp. Sta., Guthrie, Okla., *U. S. Dept. Agr.*, SCS-AP-6, p. 4, 1936.

† Should be slightly higher; clocks out of order, two rains not recorded.

The difference in the percentage of runoff, while not great, was one-sixth less from the contour planting. The loss of soil, on the other hand, was less than half as large from the contour as from the up-and-down slope planting of cotton.

In Table 23, Lamb gives some results from the U. S. Soil Erosion Experiment Station in southern New York for 1935.

The rainfall during the period covered (July 7 to Nov. 15) was 17.94 inches. The plots were 311.2 feet in length and 21 feet in width; the slope of the upper half was 7 per cent and on the lower half, 14 per cent. In July the rainfall was 5.53 inches, with 0.60 inch as the highest precipitation in 10 minutes; in August, 4.69 and 0.55 inches; in September, 2.94 and 0.12; and in November, 3.57

inches with 0.46 inch the maximum in 10 minutes. From the data in Table 23, it is evident that the loss of water was twenty and of soil more than one hundred times as great from the up-and-down hill potato rows as from the contour rows. The saving of soil was particularly notable.

TABLE 23.—EFFECT OF CONTOUR AND UP-AND-DOWN HILL PLANTING OF POTATOES, U. S. SOIL EROSION EXPERIMENT STATION, NEAR ITHACA, N. Y.*
(Upper half of plots slope, 7 per cent; lower half, 14 per cent)

Plot	Direction of rows	Loss of water, per cent	Loss of soil per acre, pounds
K3	Up-and-down hill	13.64	27,757
K4	On the contour	0.71	202

* LAMB, JOHN, JR., U. S. Erosion Experiment Station Results for 1935, *Soil Conservation News*, Ithaca, N. Y., pp. 5–8, February, 1936.

Nichols reports results of contour cropping in Alabama.[1] On slopes of 10 per cent, contour cropping held 1 inch of water which was applied to saturated soil in 9 minutes with but slight loss of soil. Under the same conditions but with the crop rows running up- and-down hill, nearly a ton of soil was lost from each acre as a result of that application of water. On slopes up to 20 per cent, the loss of soil was nearly 60 tons an acre with the rows running up- and-down hill. Even with contour cropping the loss increased rapidly on slopes steeper than 10 per cent.

Performing all tillage, seeding, and harvesting operations on the contour aids greatly in the control of erosion.

Fruits on the Contour.—The planting of all fruits on the contour is essential on all erosive lands. Many soil-conservation principles apply to fruit the same as to other crops.

Strawberries.—Strawberries, which are set annually, or at least frequently, may readily be planted on contour lines when a bed is being reestablished. If set with the rows running up and down the slope, erosion accompanies heavy rains the first year. During subsequent years, however, erosion is not likely to be serious, owing to the soil's being protected by the strawberry plants and the mulch usually used.

[1] NICHOLS, M. L., *Alabama Agr. Exp. Sta., Ann. Rept.*, pp. 12, 13, 1931.

Bush Fruits.—Bush fruits, being relatively short-lived, need not be disturbed even on steep slopes if proper measures are taken for the control of soil erosion. When renewal is required, however, all bush fruits should be planted on the contour. Cultivating may then be done on the level, and soil movement be kept under control.

Grapes.—Grapes present a difficult soil-erosion problem because suitable climate is such an important limiting element in grape production. Some of the areas that have the most suitable grape

Fig. 62.—Grapes on the Lake Erie plain. The slope in these rows is 2 per cent, yet some washing has taken place on this lake-laid soil. Washing is rather severe on slopes of 5 to 10 per cent.

climate are located on steep slopes facing lakes. In some of these hilly vineyards, grape plants generally were set with little regard for slope. The result is that in the Finger Lakes Region of New York the vineyards, in which the grape rows run up- and downhill, are badly eroded; in fact, some vineyards have been abandoned because the loss of surface soil has rendered them unprofitable (Figs. 62 and 63).

In comparatively old vineyards, which, although badly eroded, still retain some of the original surface soil it may be wise to remove some of the present old vines and set new ones in rows across the slope on the level. Only a part of the vineyard area should be renewed each year in order not to disrupt production plans or reduce income too drastically during any one year. Fortunately, grapes come into bearing in a comparatively few years. The extreme upper part of the slope may be reset and cultivated on the contour the first year. This decreases the quantity of water flowing down the slope between the old up-and-down hill rows, and erosion on the lower part of the slopes is reduced accordingly. By planting new vines in a strip across the slope immediately below the strip set the previous year, the

FIG. 63.—Grapes on the contour, an excellent practice. These grapes are growing on a slope of 31 per cent yet owing to their being planted on the contour little damage was done even by the heavy downpour of July, 1935. Washing was severe in the up-and-down hill vineyards near by.

entire steeply sloping vineyard area may be renewed in a period of five, ten, or fifteen years as desired or as conditions may warrant. Furthermore, renewal in this manner affords an apportunity for shifting from present varieties to possibly more productive and more profitable ones (Figs. 64 and 65).

In some vineyards the present surface consists of a rock pavement without any fine soil material. The original surface 6 inches or more of soil has been allowed to wash away during the years of clean cultivation. A self-seeding legume cover crop might be used to reduce runoff and erosion, but it would compete with the grapes for moisture and plant nutrients; the latter,

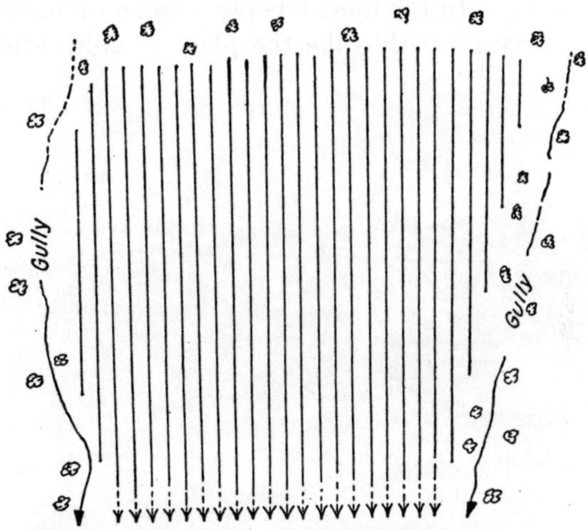

FIG. 64.—Sketch of up-and-down hill grape rows.

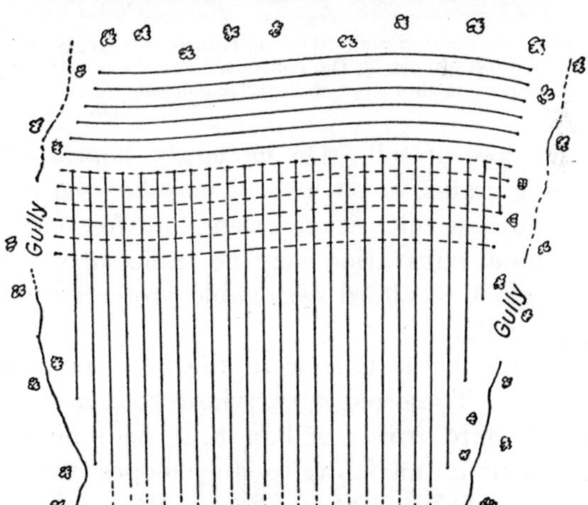

FIG. 65.—Changing the direction of rows in a vineyard. Here a start is being made in changing the rows from up- and downhill to the contour. The vines are removed from the upper end of the original rows and new vines planted on the contour the first year. This operation is repeated immediately below, the following year and so on until the vineyard is on the contour. After this has been done the vineyard with good care may be relatively permanent.

however, might be supplied economically in purchased fertilizer materials.

Orchards.—In bearing orchards, any change in the direction of cultivation is difficult because that is determined by the location of the trees. In the most steeply sloping orchards, sod with fertilization may properly take the place of cultivation in order

FIG. 66*a*.—Contour-planted orchard in the Hudson Valley, New York. These apple trees are planted almost on the contour. Alfalfa is being grown between the trees. Soil erosion is thus kept well under control.

to reduce the loss of soil. On moderate slopes, self-renewing legumes, spring grains, or buckwheat may be used under certain conditions, and less cultivation given than formerly. In some orchards diagonal cultivation may be sufficiently near at right angles to the slope to afford reasonable protection from erosion (Figs. 66*a* and 66*b*).

Many orchards in the East were killed by the severe cold of February, 1934. These old sites, whose soils are well adapted for growing orchard fruits, will in time be replanted. When replanting is done on moderate to steep slopes, the trees should be placed on definite contour lines. When desirable, such orchards can then be cultivated without inducing erosion, as cultivation has done in many old orchards.

Peach Orchards.—In some erosive peach-growing areas in the South the trees are planted on low terraces. In other areas the orchard is terraced as required for erosion control, and the trees planted in contour rows. Cultivation may then be done parallel to the terraces which are maintained by the cultivation.

Contour Furrows.—The contour furrow, or hillside ditch, is rapidly coming into favor in different sections of the country. The term ditch is usually associated with the disposal of water; the contour furrow, on the other hand, is used for the purpose of

FIG. 66*b*.—Contour-planted orchard.

holding or conserving water. Contour furrow describes this device accurately and is, therefore, used in preference to hillside ditch in this discussion. Either the contour furrow or the hillside or diversion ditch may be used to advantage on sloping fruit lands.

CHAPTER X

CONTOUR STRIP CROPPING

For some years an occasional farmer has plowed and planted moderately steep hills in strips across slopes, but these strips have not always been on the contour (Figs. 67 and 68). Some strips have been wide, and on some strips the rows have sloped

Fig. 67.—Early strip cropping in Pennsylvania. It is evident that strip cropping was not adopted until after erosion had become severe. Even so it was a commendable beginning of a good practice. Photographed, 1924.

toward one end; consequently some erosion has taken place. Even so, this early strip cropping was a most commendable effort, and from it the present strip-cropping system of erosion control has developed. The term *strip cropping* means the growing of crops in narrow strips run across slopes on the contour.

For strip cropping, slopes are divided into relatively narrow strips laid out on the contour across the slope. This method of soil utilization displaces the plan of plowing the whole of a long steep slope for one erosive crop at one time which always invites serious erosion.

The strips are planted to a rotation of crops so arranged that grasses and other close-growing crops are always placed between strips of clean-tilled crops. Ordinarily, the close-growing crops

FIG. 68.—A farmer's strip cropping in southern New York. A very excellent beginning of strip cropping. Photographed, 1934.

FIG. 69.—Strip cropping in Steuben County, New York. These strips were laid out on the farm of a cooperator by the U. S. Soil Conservation Service, 1936. Note the length of the strips. Plowing, seeding and tillage, in time, tend to eliminate gullying. (*Courtesy of U. S. Soil Conservation Service, New York.*)

take up the rain that falls on them, and any runoff from the clean-tilled strips is caught by the close-growing crops. Most of the soil washed from the strip above is held, and much of the water soaks into the soil and thus is saved from runoff (Fig. 69).

Contour strip cropping has many advantages over up-and-down hill farming. One of these is that the work of plowing, seedbed preparation, seeding, cultivation, and harvesting is done across the slope. This is easier on teams than is up-and-down hill work. Moreover, contour tillage keeps the surface of the soil smooth across the slope. Depressions are filled, and thus the beginning of gullies is very largely avoided. All of the advantages of contour farming are enjoyed in contour strip farming (Fig. 70).

FIG. 70.—Strip cropping near High Point, N. C. The strip-crop rotation on farm of a cooperator was laid out by the U. S. Soil Conservation Service. Rotation, corn, oats, and lespedeza. Photographed, 1936.

Information Needed for Developing a Strip-cropping System.—In order to work out a suitable system of strip cropping for the control of erosion the soil conservationist needs (1) an intimate knowledge of the agriculture of the area in which he is working; (2) familiarity with whatever rainfall data are available that have a bearing on the local situation, particularly any detailed intensity-frequency precipitation data; (3) a fundamental knowledge of soils, in general, and of the physical characteristics of the soils of the area, in particular, especially with reference to relative speed of percolation of water into the soil, its depth, and its total water-holding capacity; broad training in soils, some years of mapping soils being highly desirable; (4) an accurate, relatively large-scale soil map of the individual farm with field boundaries; (5) a map of the same scale showing the percentage of slope and the degree to which erosion already has taken place on the farm; (6) another map of the farm showing field boundaries and the crops being grown or that were grown the previous season, depending upon the time of the year; (7) the relationship of the different crops to soil-erosion control; and (8) the acreage of each crop grown the previous year or being grown and the farmer's crop-acreage plans for the future, based upon feed- and cash-crop requirements. The proposed strip-cropping layout can be developed on the crops map. The crop divisions and field boundaries are helpful for keeping one's location and to the farmer for visualizing the proposed changes.

Attention has already been called to variations in the erodibility of soils, (page 66) and these require full consideration in the determination of the proper widths of strips to use. With all of this information the soil conservationist is prepared to enter into a discussion of the erosion problems and plans for solving them with the farmer.

Width of Strips.—Additional experimental work is badly needed on the width of strips that may be expected to protect the soil against destructive erosion. The best width of strip for any area must necessarily vary with the steepness of slope, with the rapidity of percolation of water into the soil, and with its total absorptive capacity, and with the rainfall, with respect both to intensity and to total precipitation during periods of several consecutive days. Many years will be required to develop data from which definite conclusions can be drawn as to the specific or exact width of strip that will be most effective in the control of erosion under varying conditions.

During April, May, and June, 1935, the writer was associated with the U. S. Soil Conservation Service in southern New York. In the absence of recorded experiences with strips of different width a formula had to be developed. As a result of intensive observation of erosion conditions in the area and with an intimate knowledge of the soils of the area gained in working on these soils during the past decade, and more particularly the previous year when the writer sampled every soil type in the county, the following was proposed provisionally by him. After full discussion the formula was adopted, subject to change upon the accumulation of applicable information.

In the area were three distinct soil conditions with respect to drainage through the soil:

1. Well-drained soils, represented by the Bath, Lordstown, and Wooster series; and the Groton with good to excessive drainage.

2. Intermediate to poorly drained soils, represented by the Mardin series.

3. Poorly drained soils, represented by the Fremont and Volusia series.

The difference in the drainage of groups 2 and 3 lies largely in the difference in the depth to the impervious layer. This layer is not far below the plow line in many fields of Fremont and Volusia soils, particularly if moderate to severe sheet erosion has taken place.

A slope of 15 per cent was determined upon as a base, and 100 feet as the width of strip for the soils in group 1, 75 feet for the soils in group 2, and 50 feet for those in group 3. The width of strip was to be varied 5 feet for each change in slope of 1 per cent either toward "leveler" or toward steeper land. The way this formula worked out for different kinds of soils and on varied percentages of slope is shown in Table 24.

TABLE 24.—PROVISIONAL STRIP WIDTHS FOR SOILS WITH DIFFERENT SLOPES AND VARYING DRAINAGE THROUGH THE SOIL

Slope, per cent	Soil groups, feet		
	No. 1, well drained	No. 2, medium-poor drainage	No. 3, poor drainage
5	150	125	100
10	125	100	75
12	115	90	65
15	*100*	*75*	*50*
16	95	70	
17	90	65	
18	85	60	
20	75	50	

On slopes of less than 10 per cent, particularly those approaching 5 per cent, the width may safely be increased somewhat more rapidly than the formula calls for, but the acreage in the southern New York area with slopes of 5 per cent or less is small and consequently required little consideration. Moreover, if true-contour farming is practiced on the soils of group 1 in this area, erosion is kept under fairly good control on slopes of 5 per cent and less. Strips of potatoes and corn laid out on this basis came through the heavy prolonged rains of the summer of 1935 and the erosive period of the fall, winter, and spring following in a satisfactory manner.

After a year of experience with, and observation on, the results of strip cropping in southern New York, the Conservation Service has adopted the widths set forth in Table 25 as a guide to the various workers in developing strip-cropping systems on farms of the area.

TABLE 25.—STRIP WIDTHS USED ON NEW YORK SOIL-EROSION CONTROL DEMONSTRATION, PROJECT 1*

Slope, per cent	No. 1, well drained, feet	No. 2, medium-poor drainage, feet	Slope, per cent	No. 3, poor drainage, feet
0 to 7½	200	150	0–4	100
7½ to 12½	150	100	5–9	75
12½ to 17½	100	75	9–14†	50‡
17½ to 22½	75	50‡		
22½ to 25†	50			

* Data supplied by H. F. Eaton and W. W. Reitz.
† Slopes steeper than these are not recommended for strip cropping in this area.
‡ Strips narrower than 50 feet not recommended.

For soils such as those in group 3 the width can probably be safely increased to 100 and 75 feet, respectively, on the very moderate slopes indicated in the table. The soils in group 3 wash very easily when saturated, and, owing to their low absorptive capacity and slow drainage through the subsurface stratum, they are saturated throughout the spring and following prolonged rains. For these reasons strips on these impervious soils must not be of great width.

Attention is called to the fact that strip cropping is not recommended on slopes steeper than 25 per cent on well-drained soils or 22½ per cent on medium-poorly drained ones or on slopes steeper than 14 per cent on poorly drained soils. No strips should be much less than 50 feet in width owing to inconvenience in tillage and harvesting.

Strips of the widths stated for southern New York are entirely too wide for effective control of erosion in some sections, particularly in the South. On the pervious soils of the Midwest, however, the widths stated in Table 25 for the soils of group 1 may be suggestive until specific widths can be worked out based on the prevailing soil, crop, slope, rainfall, and erosion conditions.

In discussing "strip farming," Roe and Neal[1] write that the strips of cultivated land can range from 50 to 100 feet, depending on the steepness of slope.

Carrier and Kell, in discussing "strip cropping," say that widths of 100 to 250 feet are often advised on slopes up to 10 per cent.[2] On slopes over 10 per cent and on more erodible soils the strips should be reduced in width to a minimum of 50 feet.

The effects of different widths of strips on losses of water and soil at Tyler, Tex., reported by Hendrickson and Baird, are given in Table 26.[3]

Smooth culture was practiced without fertilizer treatment for the cotton, and other conditions were similar on all the cotton plots. Bermuda-grass (*Cynodon dactylon*) sod is given for comparison on the 16.5-per-cent slope on the Kirvin soil.

Under the conditions obtaining at Tyler the standard strip width of 72.8 feet is too wide for moderately good control of soil losses if dependence is placed on strip cropping alone. The narrow strip, 36.4 feet in width, however, is somewhat narrow for practical field operations. The wide strip, 145.6 feet, is entirely too wide for conditions at Tyler, Tex., because the loss of soil was almost double that from the standard width of strip. Curiously enough, the percentage of runoff is nearly the same on all three strip widths.

The effect of Bermuda grass on both soil and water losses is remarkable in comparison with the losses from similar soils growing cotton.

On the 16.5-per-cent slope on the Kirvin soil the strip of standard width lost one-half more water and almost twice as

[1] Roe, H. B., and J. H. Neal, Soil Erosion Control, *Minnesota Agr. Ext., Spec. Bull.* 170, p. 16, 1935.

[2] Carrier, Lyman, and W. V. Kell, Strip Cropping, *U. S. Dept. Agr., Soil Conservation Service*, SCS-TP-2, p. 4, November, 1935.

[3] Hendrickson, B. H., and R. W. Baird, Summary of Principal Results, Soil Erosion Experiments, Tyler, Tex., *U. S. Dept. Agr.*, SCS-EP-3, pp. 14, 15, 1935.

much soil as did the narrow strip. Even the narrow strip lost slightly more soil from the 16.5-per-cent slope than did the standard width of strip from the 8.75-per-cent slope. The loss of 20.20 tons of soil annually from the narrow strip on the steep slope would result in the removal of the entire top 7 inches of

TABLE 26.—EFFECTS OF WIDTH OF STRIPS OF CONTINUOUS COTTON ON KIRVIN AND NACOGDOCHES SOILS, TYLER, TEX.*

Width of strips, feet	Rainfall lost as runoff, per cent	Soil lost per acre, tons
Kirvin soil, slope 8.75 per cent, four-year summary, 1931–1934, average rainfall, 40.52 inches		
Narrow,† 36.4	16.71	12.98
Standard, 72.8	19.52	19.08
Wide, 145.6	18.69	35.98
Kirvin soil, slope 16.5 per cent, three-year summary, 1932–1934, average rainfall, 44.97 inches		
Narrow, 36.4	8.24	20.20
Standard, 72.8	12.15	34.07
Standard, 72.8	0.30	0.0065‡
Nacogdoches soil, slope 10 per cent, three-year summary, average rainfall 42.60 inches		
Narrow, 36.4	16.23	5.49
Standard, 72.8	16.76	6.45

* HENDRICKSON, B. H., and R. W. BAIRD, Summary of Principal Results, Soil Erosion Experiments, Tyler, Tex., *U. S. Dept. Agr.*, SCS-EP-3, pp. 14, 15, 1935.

† Applied to field conditions the terms "short" and "long" as used in Hendrickson and Baird's tables become, respectively, "narrow" and "wide" as used here.

‡ Permanent Bermuda-grass sod. Two-year summary 1933–1934; average rainfall, 43.06 inches.

soil in one hundred years. It is obvious, therefore, that growing clean-cultivated crops in narrow strips, 36.4 feet in width, without other means of protection, ruins this land which has a slope of 16.5 per cent. Such soil must be put to uses other than clean-tilled crops or given additional protection in order to avoid its being completely ruined by erosion.

Deeter reports the results of a rain of 1.09 inches in 15 minutes at Temple, Tex., on Apr. 2, 1935. Strip-cropped land alternating winter oats with cotton (bedded on Apr. 2) lost 0.48 ton of soil (dry basis) from slopes of 4 to 6 per cent. From an adjoining area bedded for cotton up and down the slope the loss of soil (dry basis) was 10.66 tons, or twenty-two times the quantity lost from the strip-cropped land. The clean-cultivated crop is alternated with winter oats, sudan grass, or redtop cane on contour strips which, Deeter states, ordinarily may vary from 75 to 100 feet in width.[1]

Similar data from other soil-erosion control experiment stations are being accumulated and, when published a few years hence, will serve as a basis for determining the best width of strips to use in various parts of the country.

Strips Best Kept on the Contour.—Strips must not vary greatly from the contour lest erosion take place between the rows of intertilled crops. The permissible degree of slope on strips varies with the erodibility of the soil. On highly erodible soils such as lake-laid ones erosion takes place between the crop rows on slopes of 1 per cent. On less easily erodible, open, absorptive soils slopes of 2 per cent in the rows may be allowed for short distances. Slopes of 2 per cent, however, should be the upper limit, and holding the grade to less than 1 per cent on uniform slopes is desirable. In clean-tilled crops such as potatoes, which are ridged during the final cultivation, cross dams may be placed about every 12 to 20 feet as needed, varying with the slope in the rows. On choppy slopes of 4 per cent, dams 4 inches high are needed every 8 feet. These dams may be made by hand or with a special attachment on the cultivator.

A Strip-crop Rotation for Southern New York and Adjacent Areas.—In a section such as southern New York and parts of adjacent states a four-year rotation is practicable in a cash-crop-livestock agriculture. Potatoes and corn are grown as the intertilled crop the first year, followed by spring grain seeded to clover and grass the second year; clover and grass for hay constitute the third; and grass, with some clover carrying over, is harvested for hay the fourth year of the rotation.

Arrangements of crops similar to those shown in Fig. 71 have proved satisfactory.

[1] DEETER, E. B., One Hard Rain Did This, *U. S. Dept. Agr., Soil Conservation,* Vol. 1, No. 1, p. 9, 1935.

This arrangement has some advantages. (1) No two adjacent strips are plowed and the soil deprived of crop protection at any one time. (2) All of the cultivated crops are in the same area, and consequently no time is lost in moving teams and implements from a strip on one part of the farm to a strip elsewhere. For

Number of strip	First year	Second year
	Area A	
1	Corn or potatoes with cover crop if possible	Spring grain seeded to clover and grass
2	Clover and grass	Grass for hay
3	Corn or potatoes	Spring grain
4	Clover and grass	Grass for hay
5	Corn or potatoes	Spring grain
6	Clover and grass	Grass for hay
	Area B	
1	Spring grain seeded to clover and grass	Clover and grass for hay
2	Grass for hay	Corn or potatoes; cover crop if possible
3	Spring grain	Clover and grass
4	Grass for hay	Corn or potatoes
5	Spring grain	Clover and grass
6	Grass for hay	Corn or potatoes

FIG. 71.—Arrangement of crops in a split rotation on strips in two areas.

example, when the cultivating, spraying, or harvesting of potatoes on strip A 1 is completed, no time is lost in driving across the end of strip A 2 in order to continue work on A 3 and A 5 in turn. In fact, work on all of the strips is done in the same way. Much of the time that has been lost in the past in going from one small lot to another on a distant part of the farm is largely saved by this strip-cropping arrangement. (3) In cutting silage corn, the binder is driven on the clover-grass stubble. There is no need to cut a row of corn by hand or to run it down with the corn binder in order to get started harvesting the corn strips. (4) In harvesting potatoes much driving may be done on the clover stubble in order not to leave open up-and-down hill wheel tracks on the potato land after the crop is harvested. The same holds in a measure for the corn land, which, however, may erode somewhat less seriously after harvesting than does potato land. (5) On area B in the first year the grain binder is driven on the meadow; consequently, no grain is run down and lost in getting the harvesting started. (6) Marked advantages result from harvesting the clover and grass for hay on the contour. (7) In many sections it is both necessary and desirable to pasture the aftermath or second growth on the meadow. This may be done readily because both areas, A and B, are fenced independently. Some attention may be necessary in order to avoid grazing the new seeding on strips B 1, B 2, and B 3 too closely in the fall the first year. If it is absolutely necessary to pasture the second growth of clover on strips A 2, A 4, and A 6 the first year, temporary ordinary fence or electrically charged one-wire fence may be built at slight expense in order to protect the corn or potatoes on strips A 1, A 2, and A 3.

The total number of strips on each area must be the same multiple of two in order that each crop may be grown on approximately the same acreage each year. Little difficulty is ordinarily encountered, although adjustments for this purpose are required sometimes because an occasional strip is larger or smaller than the others.

Clover and grasses are always seeded together because grass is to occupy the land the year following clover, and because a mixture of clover and grass gives better soil protection than does clover grown alone.

In contour planting the water from ordinary rains is held between the rows where it falls. In extremely heavy rains some water may break over the cultivator ridges and run off the cultivated soil onto the clover and grass strip. Here the clover and grass check the speed of the water and cause silting out of the soil particles and absorption of most of the water by the soil.

On very long slopes narrower strips may be required near the base in order to avoid erosion as a result of accumulations of unabsorbed water from the slope above. As an alternative to cutting the width of the regular strips, a buffer strip of grass may be left running through the middle of the cultivated strip. Such buffer strips of sufficient width are decidedly helpful in soil and water conservation. Twenty feet is a good minimum width.

In the area under discussion, the potato-corn year is the critical one for erosion in this rotation. The soil under the other three crops normally suffers comparatively little erosion. That being true, widening the strips beyond what is ordinarily considered

FIG. 72.—A buffer strip of meadow between intertilled crops. The beneficial effect of this buffer strip is apparent. Runoff water went through it, but much soil was held by the grass.

safe may be feasible, but only if it is planned, and the plan is carried out each year without fail, to leave a buffer strip of grass lengthwise through the potato-corn strip. In the specific situation under discussion the strip width may be increased by one-fourth to one-third if this buffer strip is left to a width of 20 feet or slightly more. This grass may be harvested for hay as usual (Fig. 72).

The buffer strip is plowed for grain very early in the spring and worked down for seeding along with the previous year's cultivated strip as a whole. This grass strip may be moved up- and downhill somewhat in order to obtain the best use of all of the land over a period of years. The use of a buffer strip makes possible widening safely all of the strips, and this in turn makes possible more of cross or diagonal harrowing than can be given on narrow strips. Diagonal harrowing helps in leveling the soil and in the control of perennial weeds.

The rotation scheme set forth in Fig. 71 is not workable for three- or five-year rotations.

On steep slopes, or if pasturing the aftermath is not required, and if loss of a little time in moving machinery from one strip to another is of no great consequence, somewhat better control of erosion may be achieved by placing all of the crops in rotation down the slope as illustrated in Fig. 73.

Number of strip	First year	Second year	Third year	Fourth year
1	Cultivated crop	Spring grain	Clover	Timothy
2	Clover	Timothy	Cultivated crop	Spring grain
3	Spring grain	Clover	Timothy	Cultivated crop
4	Timothy	Cultivated crop	Spring grain	Clover
5	Cultivated crop	Spring grain	Clover	Timothy
6	Clover	Timothy	Cultivated crop	Spring grain
7	Spring grain	Clover	Timothy	Cultivated crop
8	Timothy	Cultivated crop	Spring grain	Clover

FIG. 73.—Arrangement of crops in a regular rotation on strips.

A multiple of four strips is needed for this arrangement in order to have approximately the same acreage of each crop of a four-year rotation each year. In this system three nonerosive or close-growing crops are placed between the cultivated strips. And three nonerosive crop strips between erosive ones provide better protection than does the system illustrated in Fig. 71. Extremely heavy rains are much more likely to be entirely absorbed by three strips of clover, grass, and grain than by one of clover and grass. Furthermore, this latter arrangement is perfectly workable with a three- or five-year rotation, but fencing is required in order to pasture the aftermath.

Management of Irregular Slopes.—Owing to variations in the steepness of slopes it is impossible to keep the width of strips uniform from one end to the other and at the same time hold all of the strip boundaries definitely on the contour. Strip boundaries, however, should be laid out on the true contour or as near to it as possible.

One practical procedure is to start the plow at the lower edge of a strip and plow the major part of it. Some care is required to keep the plow turning the same width of furrow throughout the length of the strip. The plowed area should be measured or stepped carefully to determine its width. This is desirable in order that the area plowed may accommodate an even number

Fig. 74.—Management of irregular width strips. Barley and alfalfa on a slope of 4 to 6 per cent. Irregularity in width is occupied by the barley. On farm of a cooperator, near High Point, N. C.

of rows of potatoes, corn, or other intertilled crop. Thus no odd row remains to cultivate or harvest, and point rows are eliminated.

Variations in width may be absorbed in the strips that are seeded to the close-growing crops, grain, clover, grass, in the rotation previously discussed. If an unusual variation in slope causes a marked difference in width or gives a flatiron-shaped strip, the irregular-shaped correction area may be seeded to permanent meadow or to alfalfa if the soil is adapted to that crop (Fig. 74).

Management of Choppy Topography.—Field stripping, that is, cropping in strips across the main slopes, is sometimes the best that can be done. In field stripping the strips are, on the whole, not so near the contour as in strip cropping. In either case, with material slope in the rows, the placing of small cross dams between the rows of intertilled crops, as outlined on page 138, often reduces surface washing and aids in preventing the beginning of gullying.

Controlling erosion on fields that slope in several directions within short distances is difficult. Such topography is encountered in kettle and kame areas and near the heads of dendritic ravines. herever a decided dip occurs in the rows on a strip, the water flows along the rows into the depression from both directions. Unless provision is made for carrying this accumulation of water away, it causes surface erosion which is followed by gully formation. Keeping a strip of a stiff sward of grass, well adapted to the locality, prevents erosion in moderately sloping depressions of this nature. In the North the best use of such lands in general, with slopes exceeding 15 per cent on much of the area, is permanent meadow, pasture, or alfalfa under conditions favorable for it. In the South such slopes much above 10 or 12 per cent will be much safer in permanent meadow or pasture than in cotton or other cultivated crops.

Strips May Develop into Bench Terraces.—If strips are plowed downhill in exactly the same place over a period of years, they develop into bench terraces. Exactly this has happened to many fields. A pasture or woods fence runs along the edge of a sloping field which has been plowed toward the fence from year to year. Terraces from 4 to 8 feet high, which were formed in this manner, are not uncommon in hilly areas. Varying the strip boundaries first in one direction and then in the other tends to avoid, or at least to delay, the formation of an undesirable steep-faced bench terrace at the lower edge of strips. Plowing uphill instead of down on land that is not too steep aids greatly in maintaining the natural slope on strip-cropped land.

Grass Headlands Essential.—Grass headlands, or turnrows, are absolutely essential for complete erosion control in contour farming or in contour strip cropping. At the ends of strips or rows, whether these be along a fence, field boundary, road, or permanent waterway, a grassed area is needed for turning. Instead of leaving the turnrow in sod, it is often plowed, an open furrow being left at the end or along the fence. In order to utilize this land, four rows of potatoes or corn are planted across the end of the main rows. Since the land is being cropped on the contour, these end rows must run up and down the slope. If the crop rows slope toward the headland, rain water collects between the rows and flows to the headland. Such concentration of water in the headland furrow or between the headland crop rows inevitably results in the formation of a troublesome gully (Fig. 75). Because of its location, working over or alongside of such gully is difficult. If caused by the headland furrow, the gully may be so near the fence that it cannot be plowed in from

FIG. 75.—Headland gully in southern New York. The gully at the left across the grain field was caused by concentration of water in the headland of the potato field above the grain. The potato rows delivered water near the fence.

both sides or seeded easily. If left to grow, such headland gullies ultimately cause abandonment of the end of the field.

Even if the contour crop rows do not concentrate water on the headland, the headland water itself on long slopes causes appreciable erosion, especially on the lower part of such slopes. Instead of being plowed, the headland should be left permanently in sod. The sod should be fertilized, manured, and limed or all three, if necessary, for the maintenance of a thick sward. A thick-growing thrifty sward controls erosion completely on the headlands.

Grassed Waterways.—If the slopes owing to their choppy or compound nature do not admit of the working out of a practicable contour strip-cropping plan, the control of erosion can be improved by leaving the grass undisturbed in waterways or by developing a stiff sod by means of seeding, if necessary, or even by transplanting sod into the bottom of waterways. Provided the slope is not too steep, or the quantity of water to be cared for too great, a thick sward is highly effective in controlling erosion under these conditions (Figs. 76 and 77).

Grasses and Legumes Adapted for Sodded Waterways in the Northeast.—In establishing a sward for purposes of erosion control in waterways a thick-growing, turf-forming grass is best.

FIG. 76.—Grassed waterway left when plowing a pasture. Leaving this sod in the waterway is a fine idea. Greater width gives better protection against erosion. Photographed in 1912 in the writer's home county, Mercer, in western Illinois.

The fine-growing, flexible ones, or those which bend easily under the pressure of flowing water, are most desirable. In the northeastern states, for example, the fine flexible grasses, such as Kentucky bluegrass (*Poa pratensis*) and Rhode Island bent grass (*Agrostis tenuis*) and fescues, such as the creeping red fescue (*Festuca rubra* var. *arenaria*), are well adapted for use in waterways that are dry except immediately following rains. Usually sufficient nutrients are washed in from the adjacent soil to keep the grass in waterways in a thrifty condition. If, however, the slopes from which the wash comes are badly eroded, seeding legumes, such as yellow trefoil (*Medicago lupulina*) and bird's-foot trefoil (*Lotus corniculatus*), are helpful. These hardy, long-lived, fine-stemmed legumes fix nitrogen, which aids in maintaining a strong erosion-resisting sward in the waterways. If the grasses were kept down sufficiently by cutting, wild white clover (*Trifolium repens*, var.) might be used; but if the grasses are to be cut for hay or are left uncut, the trefoils or some other legume function better than does wild white clover.

Rough-stalked meadow grass (*Poa trivialis*) is well adapted to shady or moist-to-wet places. It grows well under normal moisture conditions but is sensitive to drought. This grass

FIG. 77.—Grassed waterway in west central Illinois. Slope of waterway 2 per cent, from right 7 per cent, from left 8 per cent. Photographed, 1936.

spreads by means of decumbent stems which take root at the nodes. Owing to its habit of growth, rough-stalked meadow grass produces a thick-matted sward which is ideal for protecting waterways against scouring. Best results follow seeding a mixture of grasses and legumes. If conditions at one spot or at one season are unfavorable for one species, another usually makes the necessary growth to give the desired protection against erosion.

For Moist Situations.—Creeping bent (*Agrostis palustris* or *Agrostis stolonifera* var. *compacta*) or redtop (*Agrostis alba*) and alsike clover (*Trifolium hybridum*) make good growth in situations that are wet in the spring and remain moist between rains. These grasses are not sensitive to fairly high soil acidity. Rough-stalked meadow grass, also, thrives under these conditions, but it is sensitive to relatively high soil acidity. Canada bluegrass (*Poa compressa*) is useful in places that are wet in the spring and become bone dry in midsummer. Canada bluegrass grown alone, however, does not make a thick sward capable of giving complete soil protection. It grows well on a wide range of comparatively low fertility levels. Canada bluegrass is useful in mixture with other grasses, particularly on heavy soils which become very dry in summer.

Stiff Grasses Less Satisfactory than Finer Ones.—Timothy (*Phleum pratense*) and orchard grass (*Dactylis glomerata*) grow in bunches and as usually seeded do not form a continuous sward. Moreover, such stiff grasses as timothy and orchard grass on approaching maturity stand up and permit the water to flow through them instead of being overrun by it, as happens with the fine grasses. Pure stands of these stiff grasses may permit erosion between the plants. Fine grasses are needed between these coarse ones to produce complete soil protection.

Quack Grass Not Recommended.—Quack grass (*Agropyron repens*) is an ideal grass for erosion control under many conditions. Its method of growth by underground root stalks is strongly in its favor for the purpose of erosion control. Owing to its being a noxious weed throughout the area in which it grows, quack grass is not recommended for seeding in connection with erosion control.

Grasses Adapted to the Southeast.—Bermuda grass (*Cynodon dactylon*) is ideal for the control of erosion in waterways in the South and Southwest. Like quack, Bermuda grass is regarded as a bad weed by some farmers. This attitude toward it restricts, for the present, its use for erosion control in some areas. Saint Lucie and centipede grass (*Eremochloa ophiuroides*) may be more desirable in the southern part of this area. The seeding of the annual types of lespedeza and of the perennial (*Lespedeza sericea*) with grasses or alone is very helpful in the control of erosion in waterways. The annual lespedeza, however, cannot be expected to survive the competition of the grasses long in the area to which they are well adapted.

Grasses Adapted to the Southwest.—Grasses adapted for use in outlet channels in the Southwest[1] include Bermuda grass

[1] WINTERS, N. E., *Rept. Proc. Sixth Southwest Soil and Water Conservation Conference, at Tyler, Tex.*, 1935, Oklahoma Agr. and Mech. Coll., p. 29.

(*Cynodon dactylon*); *Paspalum distichum;* Dallis grass (*Paspalum dilatatum*); Saint Lucie, a variety of Bermuda grass; and woolly finger or Rhodes grass (*Chloris gayana*). Bermuda and St. Lucie grass and *Paspalum distichum,* however, are rated as best for this purpose. For use with one or more of these grasses Hendrickson and Baird[1] have seeded, experimentally, Peruvian alfalfa and *Lespedeza sericea,* Kobe and Tennessee No. 76 lespedeza (*Lespedeza striata,* var.), southern spotted bur clover (*Medicago arabica*), and California bur clover (*Medicago hispida*). Hop clover (*Trifolium agrarium*) and subterranean clover, mentioned for use in pastures, may be helpful under conditions favorable for their growth in grassed waterways. Here, as in the North, thriving legumes aid grasses materially in forming and in maintaining the thick sward desired for erosion control.

Grass for the Central Plains States.—Buffalo grass (*Buchloë dactyloides*) is useful for erosion control in waterways in western Kansas and Nebraska and in eastern Colorado. In the northern part of this area, if sufficient moisture is present, the usual northern fine-growing grasses may be used to advantage.

DIVERSION DITCHES

Diversion ditches, properly placed between strips, aid in controlling erosion by removing the water that accumulates on long slopes. During high-intensity rainfall the grass, clover, and grain strips cannot always hold all of the water until it soaks into the soil. Some of it passes down over the slope, and in extremely heavy showers the water accumulates to such an extent that it cuts through the sod protection on the grass strips, and large gullies result. Such gully formation is avoided by removing the excess water through diversion ditches.

Diversion ditches are constructed with but slight slope and with broad bottoms in order to prevent erosion in them during heavy downpours. A fairly wide strip of not less than 20 feet of permanent meadow is required immediately above the ditch to filter out as much as possible of the soil material carried down from the land higher up on the slopes. By holding the sand and silt in the grass the effective channel of the ditch is maintained, and the risk of the water's breaking over the edge of the ditch is

[1] Hendrickson, H. B., and R. W. Baird, Summary of Principal Results, Soil Erosion Experiments, Tyler, Tex., *U. S. Dept. Agr.*, SCS-EP-3, p. 10, 1935.

reduced. If bad breaks occur in the dike or side of the ditch, the danger of destructive gully formation is greater than if the water were not thus concentrated in a ditch. Careful, intelligent maintenance of an effective channel, therefore, is an added duty for the farmer, once a diversion ditch has been made (Figs. 78 and 79).

An important advantage of using diversion ditches is the improvement in the control of sheet erosion and the prevention

FIG. 78.—Diversion terrace in New York. This diversion channel is 8 feet wide and 17 inches deep with a fall of approximately 12 inches in 100 feet or 1 per cent. This ditch is expected to carry the water from 12.5 acres of land with a slope of 6 per cent immediately above the channel. This is in every sense a Nichols terrace. Details supplied by N. J. Curtis.

of gullying on the protected area. Such protection is most needed on areas growing clean-tilled crops. The water collected from above a ditch is carried away slowly and safely to a protected outlet. Diversion ditches are most helpful during heavy downpours and during excessive rainfall of long duration. If suitable outlets can be provided at moderate expense, placing a diversion channel below each fourth strip in a four-year rotation or every third in a three-year rotation may be of real service in conserving the soil.

On areas where strip cropping affords adequate erosion control, another advantage of using diversion ditches is that strip widths may be increased somewhat on the protected area immediately below the ditch. This advantage is relatively great on areas with slow-draining subsoils. Wet or seepy spots or springs located above a ditch drain into it, and the water is carried safely away. The drying up of such troublesome wet spots markedly improves the working conditions of such fields. Underground drains should, of course, be used for purely drainage purposes in all reasonably open or porous soils, because covered drains do not interfere with field operations as do open ones.

FIG. 79.—Diversion ditch between strips on strip-cropped land. In addition to intercepting the water from the slope above, "blind" ditches empty into the diversion ditch. In same area as Fig. 78. Depth of ditch 20 inches, slope 6 to 9 inches in 100 feet, area drained 12.4 acres, slope of land above, 10 to 25 per cent. Ditch fertilized and seeded to a mixture capable of resisting erosion when established. Details supplied by N. J. Curtis.

Size of Ditch Needed.—The size of the ditch required to provide the desired protection is based on a number of factors. A ditch of sufficient capacity to care for a twenty-five- or fifty-year rain looks like a canal to the layman. But protection against anything less than twenty-five-year rains is inadequate. The type of rainfall, the concentration area to be drained or the length of the slope from which the water runs into the ditch, the relative effectiveness of the crops grown in holding back rain water, and the rapidity of absorption of rainfall by the soil above the ditch all affect the size of the channel required.

Larger capacity is needed if one-half of the area above a ditch is planted to clean-tilled crops than if but one-fourth of it is growing such crops. In other words, a high proportion of meadow grasses and legumes, small grain, and other close-growing crops tends to increase absorption of water by the soil and consequently to decrease the size of ditch that is needed.

Type of Rainfall Requires Consideration.—Areas that receive high-intensity rains need larger ditches than do sections where the rain comes as relatively gentle showers.

The relative rapidity of absorption of water by the soil is important with prolonged rains but is of comparatively less consequence in areas of high-intensity rainfall. There are two reasons for this. The initial effect of a heavy shower is to beat the surface soil into a thin mud, which on flowing over the soil seals the pores. Rapid absorption of water is thus prevented. Few fine-to-medium-grained productive soils can take up the water of heavy showers as fast as it falls. Any good soil-erosion control plan must take full account of such high-intensity rains as 1 inch in 10 minutes in Alabama; 2 inches in 30 minutes or 1 inch in 10 minutes (unofficial) in southern New York; 0.35 inch in 5 minutes, 0.56 in 10 minutes, 0.76 inches in 15 minutes, 1.05 inches in 30 minutes, and 1.17 inches in 1 hour at Salt Lake City, 0.9 inch in 30 minutes and 1.01 inches in an hour at High Line, Utah,[1] and similar or greater intensities at other points in the Midwest, the Southwest, and the far West. The major part of the water from such heavy showers runs off over the surface and causes both surface washing and severe gullying on most soils that are not well protected either by forest or by a stiff dense sward.

Diversion ditches may well be made large enough to carry twenty-five-year rains without overflowing. Differences in the retarding effect of crops on the movement of water over the slope and variations in the absorption of water by the soil due to variations in tilth affect the size of ditch required. If full consideration is given to these points, a ditch with capacity sufficient for a twenty-five-year rain may well be regarded as ample for carrying away the actual runoff from a much heavier rain.

[1] Bailey, R. W., C. L. Forshing, and R. J. Becraft, Floods and Accelerated Erosion in Utah, *U. S. Dept. Agr., Misc. Pub.* 196, pp. 11, 12, 1934.

Holding down the size to absolute requirements makes possible a shallow ditch with only a low dike. A high, steep-sided dike is cut down more rapidly by erosion than is a lower, flatter sided, one. On slopes of 15 per cent, or more, the lower side of the dike is often so steep that farm machinery cannot readily be operated on or over it. The same situation obtains on the upper inside slope of deep ditches. Severe erosion takes place on such slopes before they are stabilized by grasses or other vegetation. These difficulties, in a measure, increase crop-production costs where diversion ditches are used.

Grasses for Diversion Ditches.—The most desirable grasses for use in diversion ditches are the fine-growing rather than the stiff, coarse-growing types. For the area to which they are adapted creeping bent (*Agrostis palustris* or *Agrostis stononifera,* var. *compacta*) and rough-stalked meadow grass (*Poa trivialis*) are unusually desirable. They grow in wet as well as in normally dry places. Being fine and flexible, these grasses retard the flow of water less than do the coarser grasses. Moreover, owing to their stoloniferous method of growth and fibrous root systems they afford ideal protection against scouring by the water on the bottom of ditches. In ditches that are dry between rains the addition or substitution of Kentucky bluegrass (*Poa pratensis*) and Rhode Island or colonial bent grass (*Agrostis tenuis*) and wild white clover (*Trifolium repens,* var.) improves resistance to erosion. The bent grasses with clover make better growth on a low-fertility level such as prevails in ditch-bottom subsoils than do Kentucky blue and rough-stalked meadow grass.

In any case, the growth of vegetation in diversion ditches ought to be removed by cutting two or three times a year. In irrigation ditches in the West the flexible water vegetation is mowed regularly and removed because it retards the movement of water. In other words, such vegetation reduces the capacity of irrigation ditches. So, also, the efficiency of diversion ditches is markedly reduced by coarse grasses, weeds, or shrubs in them. Furthermore, such growth causes rapid deposition of silt and sand in the ditch, which necessitates more labor for the maintenance of an effective channel than if finer vegetation is used.

In general, the grasses discussed for use in sodded waterways (pages 146 to 149) serve equally well in diversion ditches.

The slope or gradient of the ditch must be moderate if scouring is to be avoided. The type of soil material that the ditch traverses requires consideration, as does also the prospect for a satisfactory protective vegetative cover for the bottom and sides of the ditch. The steeper the grade of the ditch the more water it carries, but the erosion hazard is correspondingly increased. The grade used in terrace channels is suggestive of what is needed or what is safe in diversion ditches (see pages 176–178).

FIG. 80.—Diversion-ditch outlet. This moderately well stabilized gully serves as outlet for the ditch shown in Fig. 79. Honeysuckle (bush type) is planted on the raw banks near the stone dams. In southern New York.

Diversion-ditch Outlets.—The outlet or place of delivery or dumping of the water from diversion ditches, too, requires careful consideration. One point to bear in mind is that an old waterway which is stable under existing conditions may begin to erode if too much additional water is dumped into it. If doubt exists regarding its stability after increasing the flow of water in a gully, stones may be placed in the bottom of it, or willow stakes which take root and grow may be placed in and on the sides of such a gully. The black locust (*Robinia Pseudo-Acacia*) may be planted on the dry upper part of the gully sides, and native wild shrubs may be used as needed (Fig. 80).

Owing to the public interest in all phases of it, soil-erosion control merits a large measure of public support. This being true, much thought may well be directed to public phases of the subject, even in deciding upon the location of the outlet for a diversion ditch. If the water is spread out on a thick grass sward or in brush or woods, it is slowed down more than if it is dumped into an existing gully. Water flows slowly in diversion ditches, and this reduces the crest of floods.

One of the best places to dispose of water from diversion ditches is on heavily covered forest areas. Good judgment must be exercised so as not to dump such an excessive quantity of water in one spot as to break through the soil protection and start gullying. If water is spread out sufficiently, all of it is taken up by absorptive soils. Thus none of it flows rapidly to the streams to aggravate flood conditions. The water delivered by diversion ditches onto porous timber soils percolates into the soil, from which it reappears slowly on lower levels as springs or comes out directly into streams, or it may stay in the soil as part of the ground-water supply for wells.

The next best outlet for diversion ditches is on pastures. However, before the water is turned out it should be determined whether the sward is likely to stand the additional water without gullying. If doubt exists on this point, the grass should be manured or fertilized or both, and seed applied if necessary, a year in advance of turning out the water. Furthermore, the water must be well spread out. This may be accomplished by making narrow sodded lips in the lower edge of the ditch. The grass at the point of delivery of water from these lips is fertilized to stiffen the sward so that no gullying may start there.

Care of Diversion Ditches.—Diversion ditches require some attention. Scouring or erosion of the bottom and sides needs to be watched and should be reduced to a minimum if it cannot be prevented entirely. Rodents are likely to work in the dike and in time let the water through in their burrows. If this happens, the break must be filled at once lest a large opening be cut in the dike. Filling such breaks is expensive because it involves hand work. Grass on the bottom and sides of the ditch should be cut two or three times a year depending on the growth made. On a ditch bottom of tight-clay subsoil it is difficult to establish any vegetation whatever. In such places planting sod, even though it is expensive, may be necessary in order to provide protection.

CHAPTER XI

TERRACES

Terraces have been in use in the Orient and in Europe for many years and to some extent for more than fifty years in this country.

Benefits of Terracing.—Terraces divide a large sloping drainage area into a number of distinctly separate ones. Each area has its own drainage waterway immediately above the terrace. Previous to terracing, much water was lost, and many tons of soil were carried from the field by sheet washing and by gullying. The terrace channel, however, collects the water from its drainage area and conducts it slowly around the slope to a stabilized outlet. The terrace waterway, being broad and of low grade, the flow of shallow water is so slow as not to move large quantities of soil.

The terrace holds the water longer on the land, and more of it soaks into the soil than on unterraced land. This difference in absorption increases with the passing of time because the terraced land improves as a result of good cropping practices and the addition of organic matter. Similar unterraced land, on the other hand, deteriorates owing to heavy loss of surface soil in spite of possible additions of organic matter. Conservation of water is beneficial to crops in most seasons. Moreover, any increase in absorption is beneficial to the area because the water table has been lowered since the removal of the forests. Crop yields are higher on terraced than on unterraced land after the terraced land has attained a normal condition.

Furthermore, terraces prevent gullying, or, if moderate-sized gullies have formed, terracing fills them in a few years. Because terraces divert the rain water off the slope, the water is prevented from causing sheet erosion and gullying as it did before the slope was terraced. Many small gullies have been filled merely by terracing a slope and then farming it on the contour in a good practicable manner.

Relation of Rainfall to Terraces.—The most important variable factor affecting the size of terraces is rainfall. In planning terraces it is desirable that they be of sufficient capacity to carry any rain likely to fall on them without overtopping. For this purpose long-time rainfall records are essential. Heavy rains during both short and long periods must be provided for. Especially large capacity is required in areas of high-intensity rainfall. If one of the upper terrace embankments is inadequate and overflows, gullying on the terraces below is almost certain to follow. Moreover, the terraces below are endangered owing to the increased quantity of water that must be taken care of. In fact, destructive breaks in successive terraces below often result from a break higher up on the slope.

Relation of Soil to Terraces.—The soil requires some consideration in planning a terrace system. Coarse soils, on the one hand, absorb water both on the slope and in the terrace channel much more rapidly than do fine-grained soils, particularly if the latter are underlain by an impervious layer. Fine-grained, especially lake- or wind-laid, soils erode more easily than do coarser ones, which contain some stones. These and other soil characteristics, discussed on pages 66 to 72, have a bearing on the permissible grade and on the width and height of terrace embankment required for reasonable protection under widely varying conditions.

THE BENCH TERRACE

The early terrace in this country was of the bench type, the top being nearly horizontal, and the front or bank very steep. Vineyards planted on the contour develop into bench terraces in a period of years as a result of frequent contour tillage (Fig. 81).

The horizontal bench terrace was built on the Illinois soil-erosion control field in the extreme southern part of the state in 1906. An open furrow was maintained at the lower edge of it for drainage. Owing to variations in the width of the terraces as a result of differences in the main slope of the field, some point rows could not be strictly on the contour. These rows concentrated water in the furrow at the edge of the terrace. The soil carried in clogged the channel, and as a result the water broke over and ran down the edge of the terrace. Owing to the steepness of the face of the terrace, a little water did great damage. Furthermore, field mice and other rodents lived and burrowed in the steep, grassed face of the terrace. In renewing the terrace waterway the plow cut across the open burrows and let the water into them. Between overtopping as a result of clogging of the channel and the rodent burrows, breaks in the terrace occurred all too frequently. Such breaks were repaired by hand at considerable expense. Unless in repairing breaks the bottom of the ditch and the dike are built up well above the necessary final height, settling takes place to such an extent that a low spot occupies the site of the break. If the low spot is not filled in further after settling and before the next heavy rain, the break is torn wide open again.

FIG. 81.—A bench terrace. This bench terrace has developed during years of cropping terraced land. In Anson County, North Carolina. (*Courtesy of U. S. Soil Conservation Service, North Carolina.*)

In spite of these difficulties, the bench terraces in the Illinois experiment conserved the soil and increased crop yields in comparison with similar unterraced land on the same farm.

The entire area had undergone severe sheet erosion, and some gullying had occurred previous to the time when it was acquired by the state for experimental purposes.

Experimental work was conducted on four separate areas, each of which was divided into a series of plots. These series were designated by letters A, B, C, and D. Series A was protected by means of horizontal bench terraces placed at vertical intervals of 5 feet. On series B, embankments and diversion ditches gave an effect somewhat similar to that of the broad-base terrace. On series C, all tillage and planting were done on the contour. Organic matter, including 8 tons of manure to the acre for corn, except during two years, was turned under, and deep contour plowing, planting, and cultivation were practiced but with no additional erosion-control measures. Series D, the check, was farmed in the most convenient manner without special effort to prevent soil erosion. The yields obtained over a period of ten years are given in Table 27.

TABLE 27.—AVERAGE OF COMPARABLE YIELDS PER ACRE, VIENNA FIELD, ILLINOIS, (1906–1915)*

Crop	Years	Series			
		A, horizontal bench terrace	B, embankments and diversion channels	C, organic matter and contour tillage	D, check
Corn, bushels....	7	31.4	32.4	27.9	14.1
Wheat, bushels..	7	9.0	12.7	11.7	4.6
Clover, tons.....	3	0.68	0.97	0.80	0.21

* MOSIER, J. G., and A. F. GUSTAFSON, Washing of Soils and Methods of Prevention, *Illinois Agr. Exp. Sta., Bull.* 207, p. 547, 1918.

Close approximation to uniformity of the plots of the different series was not possible because of gullying and differences in slope. The best comparison from every standpoint is between series C and D. From the data it is apparent that series C, owing to the adding of organic matter and the practicing of contour plowing, seeding, and cultivation, produced markedly higher yields than did the check series. In fact, contour tillage and seeding together with the addition of organic matter on series C made a remarkably good showing in comparison with the terraced series A and B.

Owing to the frequent breaks in the horizontal bench terraces under the climatic, soil, and cropping conditions prevailing in extreme southern Illinois, these terraces were abandoned after ten years. Experimental, deep contour plowing, contour tillage, and seeding, together with the maximum, practicable return of organic matter to the soil were continued on the less steeply sloping parts of the field. The steepest land was seeded to alfalfa.

Under other climatic, soil, and cropping conditions in other parts of the country, the horizontal bench terrace has given good results. For all practical purposes, the bench terrace has been displaced by other types of terraces. The level terrace, however, may be revived for strictly permanent agriculture in intensive production on moderately steep slopes.

FIG. 82.—The original Mangum terrace. This terrace was built by Priestly H. Mangum, near Wake Forest, N. C., a full half century ago and is still serving its original purpose. P. H. Mangum, son of the builder, is standing on the crest of the terrace ridge. Last year's cotton stalks had not been plowed under. Photographed by the writer, May 5, 1936.

THE MANGUM TERRACE

In March, 1885, Priestly H. Mangum laid out and built his first broad-base terrace on his farm near Wake Forest, N. C.[1] The broad-base terrace, strictly speaking, consists of a broad, low embankment and a diversion channel. Mangum laid out his terrace with a fall of 1½ inches in 14 feet, or about 10.5 inches in 100 feet. The grade that he used was much steeper than that commonly employed today. Nevertheless, the original Mangum

[1] Information supplied by P. H. Mangum, son of the originator of the Mangum terrace, in a personal letter to the writer, dated Apr. 7, 1936.

FIG. 83.—Another terrace on the Mangum farm. This gives an indication of how well these broad-base terraces, originated by Priestly H. Mangum, have been preserved. Note the broad water channel on the upper side of the ridge on which P. H. Mangum is standing. Located near the terrace shown in Fig. 82. Photographed May 5, 1936, near Wake Forest, N. C.

FIG. 84.—Newly made terraces. These terraces were made in the spring of 1936 on the demonstration project near High Point, N. C. Photograph taken while in the company of C. C. Abernathy of the North Carolina project of the U. S. Soil Conservation Service.

terrace (Figs. 82 and 83) has controlled erosion reasonably well over a period of fifty-one years. Coarse material in the surface soil evidently protected the terrace channel from excessive scouring. In building the dike of the Mangum terrace essentially an equal quantity of soil was taken from each side (Fig. 84).

THE NICHOLS TERRACE

In building the Nichols terrace nearly all of the soil (all, if a blade terracer is used) is taken from the upper side and moved downhill to form the terrace embankment. Part of the channel, since the soil out of it is used for the embankment, is below the original surface of the soil. The lower side of the embankment need not be very steep because little or no soil is removed from that side.

Several advantages are claimed for the Nichols as compared with the Mangum terrace. (1) Less power is required to move the soil down- than uphill. (2) The outer side of the embankment is not so steep. It is easier to use large implements on it. (3) Fewer failures occur because part of the water channel is below the original surface of the soil. If the entire embankment is cut out, some of the water channel remains, and less repair is required. In the Mangum terrace a cut in the dike is much deeper because it was built up partly from below; consequently, the dike has less support on the lower side and is more vulnerable to erosion.

Both the Mangum and the Nichols are broad-base terraces, and both are effective aids in the control of erosion.

LEVEL TERRACES

Level terraces serve a useful purpose under certain soil and rainfall conditions. The terrace channel has no fall from one end to the other. Any soil carried down the slope is held in the channel. This soil must be removed in order to prevent undue reduction in the capacity of the channel. The terraces hold the water collected until it is taken up by the soil or until it evaporates. If the terrace ends are not closed, the water passes out slowly (Fig. 85).

The level terrace can be used advantageously in areas that have high-intensity rains and porous soils and in sections of low total rainfall but where some of the rain falls as heavy thunderstorms. Owing to the low precipitation, all of the rainfall is needed by crops, and the level terrace holds the water until it soaks into the soil. The ponding of water, however, causes inconvenience during planting and harvesting on soils that absorb water slowly. Level terraces can be used in pastures, in the same way as the contour furrow. The latter, however, gives a better distribution of the water.

Fig. 85.—Level terraces in Illinois. These level terraces were built in the spring of 1936 on the farm of a cooperator of the U. S. Soil Conservation Service near LeRoy, Ill. Soil yellow gray to yellow silt loam (timbered), slope 3 to 8 per cent, area of pasture protected, 7.5 acres. Data supplied by T. R. Wire of the Illinois project.

Ramser recommends 4½ feet of vertical space between level terraces on sand soils, 3½ feet on sandy loams, 2½ feet on clay loams, and 2 feet on clays.[1] These spacings apply to regions of comparatively high-intensity rains whose soils have high absorptive capacity.

TERRACE OUTLETS

The outlet is of first importance in planning a terrace system. If for any reason a properly protected outlet cannot be provided, the system is doomed before it is designed, and terracing under such conditions is likely to do more harm than good. The first step in developing a terrace system is an exhaustive study of the whole farm, or drainage area, to determine the right location

[1] Ramser, C. E., Farm Terracing, *U. S. Dept. Agr., Farmers' Bull.* 1669, p. 9, 1931.

for the outlets. A pasture, a wooded area, a field boundary, an abandoned road, or a stable natural watercourse may be developed into a suitable outlet (Fig. 86).

A further step is necessary. One must determine the probable effect of the water concentrated in a terrace outlet on the soil over which it must flow to the nearest stream. It is ideal to spread the water to such extent that it is taken up readily by the soil. If the subsoil is impervious, however, complete absorption

Fig. 86.—Terrace outlet in North Carolina. The baffles have been constructed and the bottom and sides are ready for seeding. (*Courtesy of U. S. Soil Conservation Service, North Carolina.*)

may be impossible. Provision must then be made for safe conduct of the water all of the way to a stream. Neglect at this point has its own critical erosion hazard.

It would be ideal for each terrace to have its own individual outlet on to a pasture or wooded area, because the quantity of water concentrated in one place by a single terrace is relatively small. If, however, terraces of maximum length dump water from both directions into the same ditch or old gully, the quantity of water is so great following heavy rains as to constitute a real problem. It is under these conditions that engineering and agronomic ingenuity is needed.

Terrace-outlet Protection.—Under favorable conditions on moderate slopes terrace outlets may be protected satisfactorily by means of vegetation. On steeper slopes, however, engineering structures are often required to afford suitable protection against rosion.

Vegetation Often Sufficient.—According to Winters,[1] terrace outlets and similar watercourses can be protected with grasses if water velocities do not exceed 5 feet a second (equivalent to more than 3 miles an hour). Erosion occurred, however, in all cases with a velocity of 9 feet a second. On outlets with gentle

FIG. 87.—Newly sodded terrace outlet. This outlet is protected with sod. Slope, 8 per cent; some silting has occurred. Data supplied by C. C. Abernathy of the Soil Conservation Service in North Carolina. Located near High Point.

slopes the fine-growing grasses already mentioned afford the best protection against erosion. On steeply sloping outlets, on the other hand, retarding the flow is desired. For this purpose the stiff, coarse-growing grasses serve best. Winters states that Bermuda grass (*Cynodon dactylon*), *Paspalum distichum,* and Buffalo grass (*Buchloë dactyloides*) are used, but Bermuda grass and *Paspalum distichum* give the best control. Italian rye grass (*Lolium multiflorum*), owing to its rapid growth, is used for protection, while other grasses are making the growth necessary to afford protection to the soil (Figs. 87 and 88).

[1] WINTERS, N. E., *Rept. Proc. Sixth Southwest Soil and Water Conservation Conference, Oklahoma Agr. and Mech. Coll.*, p. 29, 1935.

Sod may be laid in strips across the bottom of outlets, covering one-fifth to one-third of their area. The bare space is seeded to a mixture of adapted grasses and legumes. According to Winters, sod protection of outlets for areas up to 6 or 7 acres is limited to "slopes not exceeding 12 per cent." It is desirable to limit the water velocity to 5 feet a second. Outlets for this type of situation should be from 8 to 16 feet in width.

Structures Sometimes Needed.—Wooden spreader boards may be used, if needed only temporarily; for permanent use, however,

FIG. 88.—Meadow strip serving as terrace outlet. The grass may be harvested for hay as usual. An excellent type of protection. Large quantities of water can be carried without erosion. (*Courtesy of U. S. Soil Conservation Service, North Carolina.*)

concrete or stone spreaders set flush with the bottom of the channel aid grasses in protecting the soil against erosion.

It is where deep, old water courses, old roads, or gullies are used as terrace outlets that slopes are often so steep that grasses fail to prevent erosion. In such situations slopes must be paved with stone or concrete, or check dams built of logs, stone, or concrete. Planting shrubs at the ends of such structures often helps the structures to serve their purpose more fully than is possible otherwise (Fig. 89).

Pasture.—A pasture with a thick, thrifty sward furnishes an excellent outlet for the water from a terrace or a terrace system. Advance attention to the condition of the pasture sward at the point of the outlet is essential. If the sward consists of small, weak plants, gullying is almost certain to follow the dumping of water on it. Treatment of the sward with manure, fertilizers, especially phosphorus, and lime, if needed, is required for quick results. If all of the desired sward-forming plants, including a pasture legume, are not already present, seed of them should be applied. An ample area should be treated in order that the water may be spread out so thinly that the pasture sward can carry it without any erosion.

FIG. 89.—Coralberry improves outlet protection. Coralberry (*Symphoricarpos orbiculatus*) was planted in this terrace outlet. This shrub is a distinct aid in the stabilization of this terrace outlet. (*Courtesy of U. S. Soil Conservation Service, North Carolina.*)

If the water from terraces must be turned out without waiting a year for the sward treatment to become effective, some temporary provision must be made. If no moderately safe outlet is found, the water may be carried down the slope and released temporarily below the permanent outlet. If moderate washing occurs, that can be taken care of after the permanent outlet is established. After the desired improvement in the sward has taken place, the water may be turned into the permanent outlet. Such fertilizer applications as may be required to maintain a safe sward outlet should be made at regular intervals.

Woods.—Terrace outlets are often made in woods to excellent advantage. Soil covered by a thrifty stand of trees and a thick mat of forest-floor litter withstands the erosive effect of a considerable quantity of water from terraces or diversion ditches The porous condition of the forest soil greatly increases its absorptive capacity.

If the forest stand is thin or open, and if doubt exists as to whether the soil can carry the added water without erosion, some planting may be necessary. As examples, black locust (*Robinia Pseudo-Acacia*),[1] Japanese honeysuckle (*Lonicera japonica Halliana*), common blackberry (*Rubus allegheniensis*), wild raspberry (*Rubus ideaus*), and many others may be planted with good results in the areas to which they are adapted. Most sections possess indigenous shrubs which may serve this purpose as well as those already mentioned do in their respective sections (Fig. 89).

If such trees or shrubs are transplanted while dormant from near-by areas with a ball of soil on their roots, they may be expected to make good growth in one season. By fall, or even immediately after planting, these shrubs aid materially in the absorption of the water and in controlling erosion.

If the water from long terraces or from several terraces is brought together, special effort to spread it may be needed. Spreading may be accomplished by cutting lips, or notches, in the embankment that forms the lower side of the outlet ditch. Part of the water is discharged at each notch and is more easily controlled than if all of it is discharged at one point. Temporary spreader boards or permanent stone or concrete spreaders may be placed perfectly level and flush with the surface of the soil. These spreaders can be used to supplement the resistance to erosion of pastures or woods.

Farm or Field Boundary.—A farm or field boundary or an abandoned road may be developed into a satisfactory terrace outlet, if neither pasture nor woods are available. Boundaries may require a year or more of treatment including the seeding of legumes and grasses for resisting erosion. A partly or fully stabilized gully may be used by breaking down the sides to form a shallow, wide, flat-bottomed channel. Seeding and treatment, as suggested above, are usually needed. Old roads may sometimes be used in very much the same way as stabilized gullies (Fig. 90).

[1] After Gray.

Natural Drainage Channel.—A natural, stable drainage channel is often available and serves well as an outlet. If large quantities of water are added to the normal flow in a drainage way, cutting may be induced on the bottom and sides of the channel. If cutting occurs, steps must be taken to strengthen the vegetative protection or to pave the bottom with stones or concrete or to

FIG. 90.—Old highway developed into a terrace outlet. This satisfactory terrace outlet has a slope of 7 per cent. Note the protection afforded by the concrete baffles. The workers on this Soil Conservation Service project at Spartanburg, S. C., feel that this outlet could be protected with sod.

erect check dams or other control structures. Sometimes planting willows (*Salix alba* or *lucida* or *vittellina*) or black locust (*Robinia Pseudo-Acacia*) a year or two ahead of turning in the water aids greatly in keeping the gully in a stable condition. Driving stakes of adapted species of willow in staggered rows across the gully helps, because the stakes take root and grow rapidly under favorable conditions (Fig. 188).

A living growing protection becomes more and more effective with the years of growth. A log or plank dam, on the other hand, decays, and destructive gullying may follow, if supplemental vegetative protection has not been established in the meantime. However, structures may properly be used to protect situations while vegetation is becoming established (Figs. 91 and 92).

Highway Ditches Not Suitable for Terrace Outlets.—Highway drainage ditches, generally speaking, are not suitable for use as terrace outlets. All other possibilities should be fully explored

Fig. 91.—Gully to be used as terrace outlet. (*Courtesy of U. S. Soil Conservation Service, North Carolina.*)

before even considering the highway ditch. If no other outlet appears feasible, the highway authorities should be consulted, and a satisfactory cooperative agreement worked out before proceeding with the laying out or installation of terraces. Only fully stabilized highway ditches are suitable outlets. Stabilization is usually accomplished by paving both the roadside ditch and the connection between the terrace and the road ditch.

Immeasurable damage has been done in the past and is still in progress in many places (Fig. 168). Terrace water is pouring over steep banks, with bad results. In places, no doubt, the road has been widened since the terraces were built, in which case the terrace outlet may have been destroyed. The fullest cooperation in these matters among the highway authorities, the farmer,

FIG. 92.—Stabilized gully serving as terrace outlet. This gully was stabilized by means of log check dams. The use of vegetation in addition will be helpful. This is the same gully as shown in Fig. 91. (*Courtesy of U. S. Soil Conservation Service, North Carolina.*)

FIG. 93.—Gully traceable to road drainage. The cause of this gully appears to have been drainage water from the highway. Each notch on the other side of the gully is the end of a terrace water channel. A well-protected terrace outlet parallel to the big gully is needed. Near Spartanburg, S. C.

and the soil-conservation and agricultural extension workers is essential for the best interests of the public (Fig. 93).

Location of Terraces.—After safe outlets have been found, the next step is a thorough joint study of the area to be terraced by the agricultural engineer and the soil man or agronomist. In fact, the location of safe terrace outlets is the concern of the engineer as much as of the agronomist. The available rainfall intensity-frequency data, the soils of the area, and the farmer's cropping system receive careful consideration. From this a joint decision is reached as to location of the terraces.

Because terraces are intended to serve for many years, and because they are built at considerable expense, the terrace system should be designed and built as nearly right as is humanly possible in the light of existing knowledge. Owing to lack of information and experience and of precision instruments in the past, some terraces have failed. The water concentrated by them has caused much gullying in some places (Fig. 96). By observing these failures, and through wide investigation and experimentation, a body of invaluable knowledge has been accumulated. In the light of this recorded knowledge and with the precision instruments in use today, the production of the "perfect" terrace is the joint responsibility of the engineer and the agronomist. The laying out of the terrace system and the supervising of its construction are the task of the experienced agricultural engineer.

Many a farm has the problem of receiving runoff from adjacent lands; in other words, the farm boundary cuts across a watershed. A single system of terraces for an entire watershed is ideal if all owners involved are ready to terrace at the same time. If no cooperative arrangement for the entire drainage area can be worked out, separate protection for the lower lying land must be provided. A diversion ditch placed near the farm boundary may be built for safe disposal of the water from the higher land. Unless a better terrace outlet is available elsewhere, the diversion channel may be used for that purpose.

Each distinct drainage area, if separated from others on the farm by deep ravines or by large natural water channels, is treated as a unit. The highest point on the area to be terraced is located, and from it the location of the uppermost terrace is determined.

Spacing Terraces.—The spacing of the other terraces down the slope is in accord with the data in Table 28 or some local modification of it known to be better adapted to existing soil, climate, or cropping conditions.

TABLE 28.—SPACING OF TERRACES IN FEET*

Slope of land, per cent	Vertical			Horizontal		
	Erodibility			Erodibility		
	Easy	Normal	Slow	Easy	Normal	Slow
Less than 1	½	1	1½			
1	1½	2	2½	150	200.0	250.0
2	2	2½	3	100	125.0	150.0
4	2½	3	3½	62.5	75.0	87.5
6	3	3½	4	50.0	58.3	66.7
8	3½	4	4½	43.7	50.0	56.2
10	4	4½	5	40.0	45.0	50.0
12	4½	5	5½	37.7	41.7	45.8
14	5	5½	6	35.7	39.3	42.9

* RAMSER, C. E., Farm Terracing, *U. S. Dept. Agr., Farmers' Bull.* 1669, p. 8, 1931.

Ramser states that the vertical distance between terraces should be reduced ½ foot from the distances given in the accompanying table, on land that is extremely susceptible to erosion, and that on land capable of absorbing much water and not easily eroded the vertical spacing should be increased ½ foot.

Based on the above notes by Ramser, the writer has added for easy comparison columns 2 and 4 in Table 28. Columns 5, 6, and 7 of the same table show the horizontal spacing or the approximate width of the terraces on the basis of Ramser's vertical spacings. Attention is called to the very rapid reduction in width as the steepness of slope increases from 2 to 6 per cent and also to the marked differences in width of terraces on the easily erodible as compared with the less easily erodible soils. On slopes steeper than 10 per cent the terraces become very narrow—in fact, too narrow for practical farming except in hay or pasture.

Martin suggests a vertical spacing of 1½ feet on 1-per-cent slopes and 5½ feet on 10-per-cent slopes as more suitable for conditions in Missouri.[1]

[1] MARTIN, G. E., Terracing to Prevent Erosion, *Missouri Agr. Ext., Circ.* 317, p. 10, 1935.

Roe and Neal suggest a vertical spacing of 2 feet on 1-per-cent slopes and 5½ feet on 10-per-cent slopes to take account of the more favorable conditions with respect to soil erosion in Minnesota as compared with other sections, notably in the southeast states.[1]

Clyde starts his table of vertical spacing with a 4-per-cent slope and 3½-feet vertical spacing and goes up to 6 feet at 14 per cent.[2] He increases the vertical spacing ½ foot for western as compared with eastern Iowa. This difference takes account of lower erodibility in the western part of the state.

Lehmann and Hanson use a vertical spacing of 3 feet and a width of 100 feet on 3-per-cent slopes and go up to 7 feet vertical spacing on 12-per-cent slopes.[3] They do not reduce the width of terraces so rapidly with increased slope as Ramser does.

Roberts, Kelley, and Welch start with a vertical drop of 4 feet on slopes of 5 per cent and go up to a 6-foot vertical drop on slopes of 10 to 12 per cent on loam soils.[4] They reduce the vertical drop between terraces 1 foot for "tight heavy soils." This is equivalent to a width of 80 feet for terraces on "loam" and 60 feet on the clay soil on 5-per-cent slopes.

Zeasman starts with a 4-per-cent slope and 3¼-feet vertical spacing on "tobacco soils and similar loose soils," 3½ feet on "loam and silt loam soils in general farm crops,"[5] and 5 feet on "sandy loam soils" in Wisconsin. On a 10-per-cent slope his spacings are 4½, 6½, and 8 feet, respectively, for the different soil and cropping conditions. He spaces terraces 9 feet apart vertically or about 75 feet wide on 12-per-cent slopes on sandy loam soils. These spacings are wider than others and reflect the favorable conditions existing in Wisconsin.

[1] Roe, H. B., and J. H. Neal, Soil Erosion Control, *Minnesota Agr. Ext., Spec. Bull.* 171, p. 4, 1935.

[2] Clyde, A. W., Terracing to Reduce Erosion, *Iowa Agr. Ext., Bull.* 172, p. 4, 1934.

[3] Lehmann, E. W., and F. P. Hanson, Saving Soil by Use of Mangum Terraces, *Illinois Agr. Exp. Sta., Circ.* 230, p. 10, 1930; also Lehmann, E. W., and R. C. Hays, Terraces to Save the Soil, *Illinois Agr. Exp. Sta., Circ.* 459, p. 12, 1936.

[4] Roberts, George, J. B. Kelley, and E. G. Welch, Soil Erosion, *Kentucky Agr. Ext., Circ.* 129, p. 19, 1922.

[5] Zeasman, O. R., Control Soil Erosion, *Wisconsin Agr. Ext., Circ.* 249, p. 16, 1931.

Carnes and Wilson space terraces on one-per-cent slopes, 2½ feet apart vertically and 180 feet horizontally in Alabama.[1] On a 5-per-cent slope they use 3½ and 75 feet respectively and on 10-per-cent slopes, 4¾ and 48 feet respectively.

Practicable Slope Limits.—In the South many regard slopes of 12 per cent as the maximum that can be terraced and farmed successfully on normally erosive soils. In Alabama, Carnes and Wilson give 43 and 40 feet as the proper width of terraces on 12.5- and 15-per-cent slopes, respectively. These terraces are rather narrow for practical agriculture. On the Mangum farm at Wake Forest, N. C., slopes of 12 to 15 per cent were terraced in an early day, but on that soil under cotton farming this steep land was allowed to grow up to pines that were 4 to 8 inches or more in diameter in the spring of 1936. It appears highly probable under present economic conditions and with present demands for farm produce that it might be wiser to keep slopes steeper than 10 per cent in more or less permanent meadow or pasture on the more erosive land. This is particularly true if this steeper land has already been washed badly. By keeping such land in a mixture of legumes and grasses the soil can be improved, and erosion controlled at the same time. If later the demand for farm products increases, this land will be in condition to produce good yields for a few years and then go back into erosion-controlling, soil-building crops again.

Height of Terraces.—Ramser in the publication referred to recommends 15 to 24 inches as the right height of the terrace embankment above the bottom of the water channel. The average for normal soil and rainfall conditions is about 18 inches, but some workers prefer 15 inches under their conditions. Some difficulties arise with terrace heights greater than 18 inches because the outer side of the embankment is so steep, particularly on slopes steeper than 10 per cent. By actual determination the slope of the outside of a Mangum terrace embankment in North Carolina was 45 per cent on a field slope of about 12 per cent. It is difficult to handle the larger harvesting implements such as the grain binder and the hay rake on the steeper terrace slopes.

Width of Terrace Embankments.—Mangum terrace embankments are built from 15 to 30 feet in width at the base. The

[1] Carnes, A., and J. B. Wilson, Terracing in Alabama, *Alabama Agr. Ext., Circ.* 148, p. 6, 1935.

broad ones have advantages over the narrow ones. Slopes of the embankment are less steep on the broad terrace; consequently, large implements can be used on them to better advantage than on narrow, steep-sloped ones. Washing is more pronounced on the narrow than on the broad terrace of the same height. More expense of maintenance, therefore, is required on the narrow terrace of normal height. Width, however, is governed in a measure by the field slope. Narrower terraces are used on the steep than on the flatter slopes.

Grade of Terraces.—The water channel of the early terraces was given more slope or grade than is regarded as safe or desirable today. In determining the grade to be given to the water channel in terracing, the soil requires consideration. The less erosive soils stand a steeper grade without washing than do the more easily erodible ones, such as the highly uniform-particled lake-laid or the fine-grained wind-laid ones. According to Ramser, 6 inches in 100 feet is the steepest grade permissible, and some soils wash even at that grade.

Terraces may be built with a uniform or with a variable grade. Those of uniform grade have the same fall per 100 feet from the upper to the lower end. The terrace of variable grade is comparatively level at the upper end, and the grade is gradually increased toward the outlet. The variable grade is more popular because of its advantages. The upper section of the terrace carries less water than do the lower sections, and more water is taken up by the soil in the section that has the slowest movement of water. Also, with less grade the water in the upper section does not run down rapidly to tax the capacity of the channel, as occurs in channels of uniform grade. Less soil is washed away in the variable- than in the uniform-graded terrace channel. With very long terraces it may be necessary to increase the terrace height toward the lower end in order to carry the water safely. An alternative is increasing the width of the channel or in decreasing the distance between terraces.

Ramser uses the terrace grades given in Table 29.

For specific soil and rainfall conditions various terrace grades are recommended. A few instances are given here from the publications by some of the writers previously cited.

Roe and Neal recommend a drop of 1.2 inches for the first 300 feet, increasing to a maximum of 4.8 inches for the section between 1,200 and 1,500 feet immediately above the outlet, for Minnesota conditions.

TABLE 29.—FALL IN VARIABLE-GRADED TERRACE*

Length of Terrace Section, Feet	Drop in 100 Feet of Length, Inches
0 to 300	½
300 to 600	1
600 to 900	2
900 to 1,200	4
1,200 to 1,500	6

* *U. S. Dept. Agr., Farmers' Bull.* 1669, p. 7, 1931.

Clyde in Iowa used a drop of 3 inches in 100 feet at the upper end and 6 inches for the last 300-foot section of a terrace 1,200 feet long.

Lehmann and Hanson use the grades recommended by Ramser on 5-per-cent slopes in Illinois, but they increase the grade materially for field slopes of 10 and 15 per cent.

Carnes and Wilson in Alabama take into account the slope of the field; the kind of subsoil, whether clayey or sandy; and, of course, the length of the terrace as well. Their recommendations are given in Table 30.

Construction of Terraces.—The construction of terraces is an engineering problem which is fully covered by Ayres[1] in "Soil Erosion and Its Control" and in publications of the U. S. Department of Agriculture. Furthermore, many of the states in which terracing is feasible have published detailed instructions for building terraces under the conditions prevailing in their respective areas. For these reasons and because of the space required to cover it in detail actual terrace construction is not given here.

The terrace at the top of the slope is laid out first and must be built first. Each terrace embankment with its accompanying water channel is planned and built with sufficient capacity to protect only the area of its own terrace. That being true, if a terrace well down on the slope is built first, it is endangered by the collection in its channel of all of the water from the unterraced slope above. In case of a heavy rain the newly built terrace is

[1] AYRES, QUINCY C., "Soil Erosion and Its Control," McGraw-Hill Book Company, Inc., New York, 1936.

overtopped, washed out, and ruined. For this reason the uppermost terrace is built first, and each terrace immediately below it in turn until the whole slope is protected by terraces (Fig. 94).

TABLE 30.—TERRACE GRADES BASED ON FIELD SLOPE, SUBSOIL, AND LENGTH OF TERRACE IN ALABAMA*

Terrace length, feet	Land slope		
	5 per cent	10 per cent	15 per cent
	Maximum fall per 100 feet, inches		
Clay subsoil			
0 to 100	0	0	0
100 to 400	1	1¼	1½
400 to 700	2	2½	2¾
700 to 1,000	3	3½	4¼
1,000 to 1,300	4	4¾	5½
1,300 to 1,600	5	6	7
Sandy subsoil			
0 to 100	0	0	0
100 to 400	¼	½	¾
400 to 700	¾	1¼	1½
700 to 1,000	1¼	1¾	2½
1,000 to 1,300	1½	2½	3¼
1,300 to 1,600	2	3	4

* CARNES, A., and J. B. WILSON, *Alabama Agr. Ext.*, *Circ.* 148, p. 7, 1935.

Terracing Gullied Slopes.—In building terraces across a gullied slope the gullies are filled first. Small ones may be filled by plowing in one or two furrows from each side. The terrace embankment can then be built up sufficiently with a shovel. Larger gullies may be plowed in the same as the smaller ones, and a dam made on which the terrace embankment is built. Such dam or gully fill may be made advantageously with a team and slip scraper. Compacting the fill as much as possible by driving the team over it reduces settling of the finished terrace embankment. Provision must be made for settling from one-third to one-half in dry, untrampled soil. It is far better to have the settled embankment somewhat higher than is absolutely necessary, than for it to settle so much that the water goes over it during the first heavy rain. If this occurs, the dam and terrace embankment are washed out, and much expense is entailed in rebuilding them. Temporarily it is advisable to give the terrace ditch slightly steeper grade immediately below than directly above the point where the terrace crosses a gully. This arrangement makes it possible for the water to get away faster than it comes from above the gully and, moreover, makes it possible for the terrace watercourse to carry away the water coming down the gully.

Cost of Terraces.—The cost of terracing is highly variable. Among the factors affecting costs are: (1) the slope of the land,

FIG. 94.—Terracing in Oklahoma. This broad-base terrace is being built with heavy terracing machinery. A few more rounds complete the ridge. (*Courtesy of U. S. Soil Conservation Service, Oklahoma.*)

more terraces being required on steep than on more gently sloping areas; (2) the kind of soil, since less power is needed for terracing a sandy loam than for a tough, heavy clay loam; (3) the length of terraces, long ones requiring less time per unit of length than short ones; (4) obstructions, because land that is covered by stones and shrubs or small trees, which must be removed, requires much time for terracing; and (5) gullies, for if gullies of a size that must be filled before terracing are present, their number and size affect costs. The skill of the operator, size and efficiency of equipment, power and labor, and the wages paid all affect costs.

Ramser gave the approximate costs per acre of constructing terraces 15 inches high, 20 feet wide, and not less than 1,000 feet long in light soils on moderate land slopes (see footnote to Table 29). For clean-cultivated land without gullies the cost was from $1.50

to $2.50 an acre; for grass or virgin land without gullies, from $2 to $3; and for clean-cultivated land with small gullies, from $3 to $6 an acre. If larger gullies or stumps and stones are present, the costs may go as high as $12 to $15 an acre.

Terraces may be constructed in less time and probably at lower cost, if everything is hired, by using large power equipment than by employing farm labor and machinery and farm horses and

FIG. 95.—Farmer-constructed terrace. This terrace, on the J. M. Dean farm three miles west of Spartanburg, was constructed with two-horse, No. 13 Oliver plow with extension moldboard and homemade, V-drag. Six rounds were made in the construction. Width of terrace, 12 feet, height 18 inches. Back furrowing had not been done when photographed. (*Courtesy of U. S. Soil Conservation Service, Spartanburg, South Carolina.*)

mules. There are, however, many excellent reasons for developing ways and means whereby the farmer may construct his own terraces. The necessary machinery may be owned by a group of farmers or by an individual and be rented to others. If the farmer wishes to do the work with his own teams, labor, and equipment during slack periods, he can then do it advantageously (Fig. 95).

As previously stated, however, the planning of the terrace system should be done by trained men. The actual laying out, supervising the building, and checking the completed system should be done by an engineer or other person with suitable experience in such work.

Care of Terraces.—Terraces require considerable attention the first year, especially if they are not compacted by the tractor wheels or by the trampling of teams during the process of building. Uneven settling of terraces results in low spots which require being built up promptly. Otherwise, water breaks over, cuts the embankment, and makes a gully down the slope. Such gullying endangers all terraces on the slope below. The first year, therefore, it is advisable to check over the terraces after every heavy rain, to build up all low spots in the embankment, and to clean out all fills in the channel.

It is better to seed the newly made terrace embankment to a close-growing rather than to a clean-tilled crop. Seeding a legume with grass or with a small-grain crop is good practice. The drill is run parallel to the embankment because this aids in reducing the washing of soil off the ridge itself.

In terracing, the surface soil is removed from the channel area, leaving it rather unproductive and subject to washing. The channel and land that had lost much of its surface soil before it was terraced is best seeded to a green-manure legume crop the first year, provided the use of the land for cash or feed crops is not absolutely necessary. Mixing a legume and a nonlegume adapted to the locality, such as grass, small-grain, or annual-hay plants, produces more organic matter for turning under and improves the control of erosion more than does seeding a nonlegume alone.

After the first year careful checking of the terraces following heavy rains and repairing the slightest damage are essential to successful protection of terraced land. Soil for patching is taken from the channel which is filled again by rains.

The sides of the terrace embankments are fairly steep; 20 to 40 per cent is not unusual. Natural forces reduce the height of embankments materially during a single year. Height is restored by throwing the furrows to the top of the ridge from both sides once a year, and twice a year may be necessary in some places. The channel is kept open by placing the dead furrow in the channel as frequently as is necessary to remove the soil washed into it (Figs. 96 and 97).

The two-way plow is very useful, if, indeed, not absolutely essential, in the maintenance of terraces and in the farming of terraced land. This implement may be used as an ordinary

Fig. 96.—Terraces which failed. Owing to faulty construction or inadequate care, or both, these terraces failed, the lower one 7 years, the middle one 5 years, and the upper one 2 years before this photograph was taken. Reclamation of this land will be costly. (*Courtesy of U. S. Soil Conservation Service, South Carolina, South Tyger project.*)

one-way plow in back furrowing at the top and on the sides of the embankment. It may be used as a two-way plow to throw the soil toward the lower embankment until the terrace channel has been plowed out. The upper part of the terrace may then be plowed uphill against the foot of the terrace ridge above. In doing this the plow is used again as a two-way implement. On land that is not too steep this method of plowing helps to keep the soil in place and to a degree, at least, neutralizes the action of

FIG. 97.—Terraces failed because of water from the highway. Water from the highway was turned into this terraced field without ample provision for it. Gullying resulted. On the Mangum farm at Wake Forest, N. C. Photographed, May 5, 1936.

gravity and the effect of soil washing. Another advantage is claimed for plowing uphill. On sod, and in a measure under other cropping conditions as well, the water coming down the land slope of the terrace is directed into the soil at the break between the furrow slices. Placing a dead furrow between the terrace channel and the upper edge of terraces may be advantageous, especially on wide terraces. This furrow holds water and brings about some percolation into the soil. Dead furrows, however, are, in general, to be avoided. The care and the maintenance of terraces and terrace outlets are so absolutely essential as to require the attention of the landowner. Unless terraces are correctly planned, built, and maintained, it were better that the land had not been terraced. Incorrect terracing or careless maintenance owing to the concentration of water by the terraces may lead to destructive gullying (Fig. 96).

Cropping Terraced Land.—Terracing is one of the more important means of controlling erosion, but terracing alone cannot provide complete protection against erosion. Rotation of crops, including the seeding of such close-growing ones as small grains, annual hay crops, and legume-grass mixtures, is no less essential on terraced land than on strip-cropped or contour-tilled

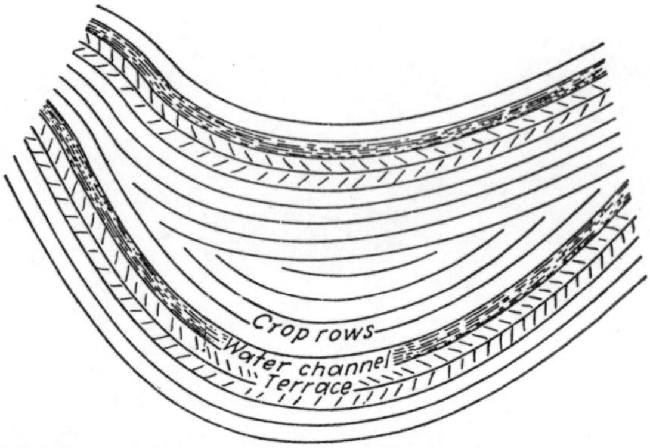

Fig. 98.—Cropping terraces of irregular widths. I. In case of any slope in the rows some washing may take place at the ends of rows in the middle of the terrace. (*Adapted from Missouri Agr. Ext. Circ.* 317, *by G. E. Martin, p.* 18, 1935.)

land. In fact, contour strip cropping on terraced land gives the best control of erosion.

Owing to variations in the steepness of slopes, terraces cannot be of the same width from end to end, if they are laid out and built in accord with the plans previously outlined (pages 172 to 179). Consequently, it is not possible to plant crops with all rows parallel to the terrace ridges and not have short or point rows.

Various ways of handling point rows are in use. For Missouri conditions, Martin suggested running the rows parallel to both the upper and lower edges of terraces with short, or point, rows in the middle of the terrace.[1] This arrangement is shown in Fig. 98. Each row holds water and encourages its absorption by the soil. He points out, however, that some concentration

[1] Martin, G. E., Terracing to Prevent Erosion, *Missouri Agr. Ext., Circ* 317, p. 18, 1935.

of water may take place where sloping point rows end in the terrace. Sheet erosion and mild gullying may result from such concentration of water during very heavy rains.

For North Carolina conditions, Holman recommends starting with rows parallel to the upper side of the terrace,[1] that is, immediately below the embankment of the terrace next higher on the slope. In this arrangement the point rows end in the terrace channel. Thus the rows discharge any water accumulated between them directly into the terrace channel at a grade not essentially different from that of the channel itself. Little silting of the channel should take place under this arrangement, which is shown in Fig. 99. Exercising considerable care in cultivating crops is desirable in order to avoid dragging much soil into the channel and thereby clogging it.

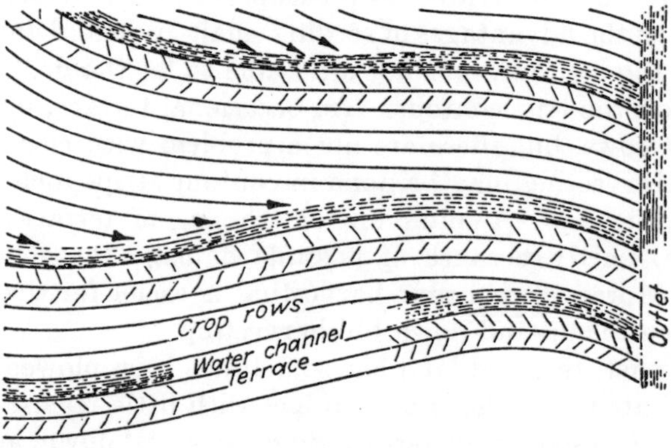

Fig. 99.—Cropping terraces of irregular widths. II. By seeding parallel to the lower side of a terrace ridge, the point rows end in the terrace channel, a very desirable arrangement. (*Adapted from North Carolina Agr. Ext. Circ.* 173, *by A. T. Holman,* p. 13, 1935.)

Planting rows parallel to the lower side of the terrace obviously places the point rows at the upper side. The water accumulated is then discharged and must flow down across the contoured rows. Some danger of sheet erosion and mild gullying exists with this arrangement of the rows.

For small-grain crops, any of these arrangements of rows is suitable, but drilling parallel to the upper side of the terrace, with the short rows ending in the terrace channel, possesses some

[1] Holman, A. T., Terracing Farm Lands, *North Carolina Agr. Ext., Circ.* 173, p. 13 revised 1935.

marked advantages. Contour tillage and seeding, however, are as essential on terraced land as elsewhere for erosion control.

Contour Strip Cropping on Terraced Land.—In some sections of the country that have a critical erosion complex, contour strip cropping is practiced on terraced land. Cotton or corn is alternated with strips of close-growing crops. In growing cultivated crops on the terrace ridges, the rows are placed parallel to the embankments. The close-growing crops check the movement of water and soil down the embankment slopes. Consequently, the ultimate loss of water and soil is markedly lowered as compared with that from unterraced slopes, a large area of which is planted to a clean-tilled crop at one time. Any irregularity in terrace width is occupied by the close-growing crop, thereby eliminating the point-row problem, which is so frequently encountered if a cultivated crop occupies the entire terrace.

Owing to the advantages of seeding small-grain and annual-hay crops on the contour, it is not advisable to let the drill rows get very far off the true contour. An occasional terrace, inevitably, has point rows, but these are not a problem with close-growing crops. Harvesting may be done in contour strips including one or more terraces. A minimum of crossing of terrace ridges is advisable, partly because of possible damage to the embankments. Moreover, contour harvesting is easier for the team or tractor than is up-and-down hill harvesting.

In the past, terraced land has sometimes been plowed, seeded, and cultivated over the terrace ridges with little regard for their location. Consequently, terrace ridges were cut down, and maintenance costs were increased. Water which collected between rows of cultivated crops carried much soil into the terrace channel and clogged it. Overtopping the terrace embankment, therefore, with its accompanying gully formation was not uncommon. On slopes of relatively slight grade, however, plowing, seeding, and other tillage operations may be done across terrace ridges. It must be fully appreciated, however, that farming terraced land in this manner necessitates more work for keeping embankments up to the required height and for maintaining the efficiency of the water channels. On the whole, however, the best control of erosion on cultivated terraced land is attained by means of contour seeding and contour tillage, even on land that has but moderate slope.

CHAPTER XII

MEADOWS, PASTURES, AND FORESTS

UTILIZATION OF MODERATELY STEEP AND STEEP SLOPES

The terms *moderately steep* and *steep* as here used in describing slopes embrace not only the actual percentage of slope but also the relative erodibility of the soil. Moderately steep to steep slopes in most sections cannot be cultivated safely. Moreover, such areas are not needed at the present time for the production of clean-tilled crops, because an abundance of food and fiber can be produced on the relatively less erodible, productive lands.

On steep slopes the strips or terraces or both must be kept very narrow in order to protect the soil from severe erosion. Cultivating steep slopes materially increases the cost of producing crops; consequently, farmers on the steeper lands cannot compete with those on the more easily tilled lands. Seeding to permanent meadow on the moderately steep land and to permanent pasture on the steeper slopes and planting forest trees on the steepest part are good utilization of such land at present. If at some future time this type of land is needed for intensive crop production, it will have been preserved. In fact, it will have been markedly improved, especially if legumes have been grown in the permanent meadows, pastures, and forests.

MEADOWS

Meadows that are classified as close-growing crops protect the soil against erosion reasonably well. The aboveground growth prevents the direct action of wind and rain on the soil, and the numerous blades and stems of grasses and legumes retard the flow of water over slopes. Furthermore, the fibrous roots hold the soil together and thereby retard both sheet erosion and gullying.

Alfalfa.—Alfalfa produces the ideal long-term meadow in the area and on the soils adapted for its growth (page 84). Grass is seeded with alfalfa for the purpose of improving the protection against erosion and for increasing the production of feed. On soils thoroughly well adapted for alfalfa, which have received the necessary lime and phosphorus, and potash, too, if needed, alfalfa may be expected to continue in production for some years. A period of six to eight years in New York is not uncommon, and some stands are known to be more than ten years old (Fig. 100).

FIG. 100.—Alfalfa and quack grass check flood water. This alfalfa has been growing on this land for ten years. In the meantime much quack grass (*Agropyron repens*) has come into it. Note the resistance of the grass and alfalfa to scouring by the flat, sharp, angular stones on this slope of from 14 to 18 per cent. No erosion took place in the alfalfa in spite of two inches of rain which fell in thirty minutes. Southern New York.

Lime sufficient for from six to eight years should be applied over a period of two or three years before seeding alfalfa in order to be certain that all of the plowed soil is in the right condition with respect to lime for this crop. Lime sufficient for so long a period is suggested because it cannot be applied effectively for alfalfa after it has been established. Regular additions of phosphorus as previously outlined aid in keeping alfalfa strong and productive.

When the stand becomes thin and no longer controls erosion effectively, it is renewed. The whole field is not plowed and reseeded at one time, because doing that invites severe erosion. If the entire field is in need of renewal at one time, alternate strips may be plowed and seeded to corn, small grain, or a summer annual, such as a millet or sudan grass.

It appears best to grow a heavy-feeding nonlegume crop in order to remove part of the nitrogen left in the soil by the alfalfa. Reseeding, however, may be done in a crop such as oats which is removed for hay. There is some danger of not obtaining a strong, long-lived, erosion-resisting stand of alfalfa in this way, however, because alfalfa requires a firm seedbed. In many areas growing a cultivated crop for a year to be followed by alfalfa is the better procedure. Alfalfa may be seeded in late spring or early summer on disked corn or potato land or on well-disked oats or wheat stubble, as is done in the Midwest where these are harvested early.

If the alfalfa stand becomes thin, and the accompanying grasses unthrifty, but the mixture still gives a fair yield of hay, it may be feasible to take from two to four years for reestablishing the alfalfa-grass mixture. In any case, not more than half of a steep alfalfa meadow should be plowed at one time and then only in strips, in order to hold soil losses to a minimum.

Long-term Meadows for the North.—Long-term meadows should always consist of a mixture of legumes and grasses. The principal limitation of permanent meadows of this type is the legume which should be perennial in character. Otherwise, the grass uses up the nitrogen left by the accompanying short-lived legume, and the yield soon diminishes rapidly along with the protection against erosion.

In the North, red clover (*Trifolium pratensis*) and alsike (*Trifolium hybridum*) serve well the first year and, under favorable conditions, carry over into the second year to some extent. For longer life, yellow trefoil (*Medicago lupulina*) and true wild white clover (*Trifolium repens*, var.) are distinctly helpful. Before seeding, conditions should be made favorable for red clover by liming, if needed, and by supplying phosphorus. Short-lived plants are included which establish themselves quickly and which are needed to protect the soil and to produce feed during the first few years while the others are getting started. Long-lived, truly perennial ones must be included to maintain the protective covering and to continue feed production for some years.

The following seed mixture,[1] suggested by Johnstone-Wallace,[2] is recommended for thorough trial in the northeast states for long-term meadows:

Seed Mixture	Pounds per Acre
Red clover (*Trifolium pratense*)	4
Alsike (*Trifolium hybridum*)	2
Yellow trefoil (*Medicago lupulina*)	2
Wild white clover (*Trifolium repens*, var.)	1
Timothy (*Phleum pratense*)	6
True perennial rye grass (*Lolium perenne*), such as Swedish Svalof Victoria, New Zealand certified permanent pasture, or American-grown seed of these varieties	4
Kentucky bluegrass (*Poa pratensis*)	2
Rough-stalked meadow grass (*Poa trivialis*)	1
Total	22

This mixture provides seeds of the plants which are depended on mainly for the first two years and also for the later years. Meadows established by seeding this mixture should be cut early, and wherever possible the aftermath closely grazed, or such meadows may be grazed after a few years.

Permanent Meadows.—If permanent, rather than long-term, meadows are desired, 8 pounds of orchard grass, preferably of a pasture type, such as the New Zealand Akaroa is recommended by Johnstone-Wallace.

Seeding these quantities of mixtures, which contain a high proportion of light seeds, such as Kentucky bluegrass, rough-stalked meadow grass, orchard grass, and rye grass, requires more than one trip over the land with ordinary seeding machines or else special seeding arrangements. A more uniform distribution of the seed is obtained if the second trip over the land is made at right angles to the first than if all of the seed is applied in one operation. A thick uniform covering of meadow plants gives the best protection against erosion.

On steep slopes on which the operation of the grain binder is difficult, the nurse-crop grain may be dispensed with. Thorough

[1] Fertilizer recommendations for New York, *Cornell Agr. Ext., Bull.* 281, p. 7, 1936.

[2] Johnstone-Wallace, D. B., agrostologist, Cornell Univ. Agr. Exp. Sta.

working including rolling is essential for success with seeding in this way. Perennial rye grass grows so rapidly as to protect the soil while the other plants are establishing themselves. Consequently, the nurse crop is not absolutely essential.

Meadow-seed mixtures should be covered from ¼ to ½ inch in depth. Failure of many seedings results from too much cover and, also, from lack of cover. The practice of seeding the meadow mixture after the grain drill in a separate operation is spreading. The seed is then covered lightly with a weeder. On the coarser soils, which are not subject to crust formation, the corrugated roller or cultipacker covers the seed satisfactorily. On steep slopes rolling on the contour is effective and aids in erosion control in comparison with up-and-down hill rolling. The latter actually encourages erosion.

Failure of a grass-legume mixture on unprotected steep land is inevitably accompanied by severe washing during heavy rains. This danger of severe erosion in case of failure of a seeding is the justification for giving the foregoing seeding details. Getting cultivated land under a vegetative cover in the least time practicable is essential for the control of erosion.

Long-term or Permanent Meadows for the South.—Permanent meadows usually are more easily attained in a natural grass climate such as that of the northeast states than in the South. Alfalfa, including the hairy Peruvian, is useful in parts of the South. Growing adapted grasses with alfalfa improves the control of erosion as already indicated. Perennial *Lespedeza sirecia* may serve much the same purpose as does alfalfa. A perennial grass with the perennial lespedeza may be used to aid in the control of erosion. The common annual lespedezas (*Lespedeza striata*) are particularly desirable and useful hay plants. Because they reseed so readily the common lespedezas serve much the same purpose as perennials.

Johnson grass (*Sorghum halepense*) and other adapted perennial grasses are highly useful both for hay production and for erosion control. The longer lived grasses should have preference.

As previously suggested for the North, if lime is needed, sufficient lime for a period of years is applied to best advantage previous to the time of establishing a long-time meadow. Phosphorus, potash, and manure may be applied at seeding time. Phosphorus, however, is needed regularly during the life of the meadow and nitrogen, too, in manure or fertilizer, if legumes are not grown regularly to supply sufficient nitrogen for the grass.

Long-term or Permanent Meadows in the West and Southwest.—In the West and Southwest the more palatable of the native prairie and other adapted grasses should be reestablished as permanent meadow for feed production and for conservation of both water and soil, especially on the land that is somewhat too steep for the production of intertilled crops. As previously indicated, any soil treatment needed aids in maintaining a grass cover capable of protecting the soil against erosion.

PASTURES

The moderately steep and erosive slopes which wash badly, if planted to intertilled crops, may be saved for feed production by establishing permanent pastures on them. In permanent pastures, as in permanent meadows, the perennial legume is the key to success. A dependable self-reseeding annual legume such as common lespedeza grown alone produces large quantities of feed and protects the soil throughout a broad east-and-west belt of states to the southward of central Illinois and southern Iowa and central Pennsylvania. In certain areas alfalfa (*Medicago sativa*) and biennial white-blossom sweet clover (*Melilotus alba*) are useful along with adapted grasses. Even though the sweet clover may reseed itself in a measure, sweet clover and alfalfa, although valuable, do not persist long enough for truly permanent pasture on erosive land.

Northern Pastures.—In the north-central and northeast states the ideal pasture legume is wild white clover (*Trifolium repens*, var.). The seed of this clover was brought to this country in hay from England by the early settlers. Soil conditions proved favorable for it, and it grew wherever the white man settled. It is related that the Indians called it the "white man's foot grass." Since those early days, wild white clover has spread throughout the entire area to which it is climatically adapted. But as the original supply of organic matter and phosphorus in the soil became depleted, wild white clover made less and less growth until it has all but disappeared from many northeastern pastures.

On very sour soils the application of some form of lime is helpful, and on sandy soils, particularly, the application of potash increases the growth of wild white clover. On many loams and silt loams in the North and Northeast, however, the application of phosphorus, alone, enables wild white clover to make excellent growth. Heavy applications of phosphorus at relatively long intervals give better results with white clover than do light annual applications. Six hundred pounds to the acre of 16-per-cent superphosphate, or its equivalent in other analyses, every three years produces a fine growth of wild white clover and grasses, if other conditions are favorable.

In the Midwest where finely ground rock phosphate is readily obtainable at low cost, it may be used in applications of from 1,000 to 2,000 pounds to the acre every four to six years. If imported basic slag comes on to the American market at a competitive price, it may well be used approximately as suggested for 16-per-cent superphosphate. Basic slag, produced in the southern iron district, is desirable for that section.

Wild white clover spreads by means of creeping stems from the joints of which roots are sent into the soil. A single, tiny weak plant often spreads over a square yard in one season after its plant-nutrient requirements have been met.

Attention has been forcefully redirected to wild white clover in this country by Prof. D. B. Johnstone-Wallace of Cornell University. He has had wide experience with this legume in both old and new pastures in Great Britain. Since 1932 he has laid down new pastures on the Cornell University Agricultural Experiment Station grounds in which wild white clover was used as the key ingredient of the seed mixture. The Kent wild white clover used was imported from England because no seed of this plant was being harvested in this country at that time.

Johnstone-Wallace's work has shown that grasses seeded alone and wild white clover seeded alone are only partly successful but that when wild white clover and the grasses are seeded together the mixture has produced excellent yields. As an average of the years 1934 and 1935, Kentucky bluegrass (*Poa pratensis*) (Fig. 101) produced 881 pounds of dry matter and wild white clover (*Trifolium repens*, var.) 3,072 pounds to the acre. On an adjacent plot a mixture of Kentucky bluegrass and wild white clover produced 4,985 pounds of dry matter to the acre (Fig. 102). The mixture on one plot produced 1,032 pounds of dry matter to the acre more than did the bluegrass and the wild white clover grown separately on two plots. Bluegrass grown alone contains

FIG. 101.—Kentucky bluegrass pasture. Kentucky bluegrass (*Poa pratensis*) when grown alone produces a thin sward which does not protect the soil against erosion unless the pasture is heavily fertilized with nitrogen. (*Courtesy of D. B. Johnstone-Wallace.*)

FIG. 102.—Kentucky bluegrass and wild white clover. The addition of true wild white clover (*Trifolium repens*, var.) to bluegrass produces a thick, erosion-resisting sward and greatly increases the production of feed. Moreover, the clover fixes nitrogen and thus enriches the soil. (*Courtesy of D. B. Johnstone-Wallace.*)

24 per cent of protein, and the bluegrass-wild white clover mixture has 31 per cent of protein in the dry matter.[1] On that basis the bluegrass produced 211 pounds of protein, and the bluegrass-wild white clover mixture 1,545 pounds of protein to the acre as an average for the years 1934 and 1935. In other words, the grass-legume mixture produced nearly six times as much dry matter and more than seven times as much protein as did bluegrass grown alone. Such decided improvement in protein content and such marked increase in yield of feed to the acre surely warrants the slight expense for wild white clover seed and the outlay for the necessary phosphorus to enable the clover to make good growth.

Proper grazing management is just as important in encouraging the growth of wild white clover as is the application of phosphorus in the production of a thick wild white clover-grass pasture sward. Neither management[2] nor the application of phosphorus alone is wholly successful. Both are essential.

Hill pastures, which produce even from one-half to two-thirds as much feed as did this Cornell pasture, are producing far more feed and are controlling erosion far more effectively than most of the untreated, unmanaged northeastern pastures are doing at the present time.

As already indicated, the pasture sward must be thick and must cover the soil completely if the pasture is to control erosion. And such a sward can be produced to best advantage only by growing legumes and grasses together. A mixed grass-legume sward, phosphated and well managed, does not heave as does either the pure grass or the pure clover sward. This is shown in Fig. 103. Far more erosion takes place during winter and spring on a soil that is badly heaved than on one on which the sward is so dense as to be lifted uniformly by the frost. Such sward settles back intact when the frost goes out. Thus neither clover nor grass roots are injured by being heaved out, and no bare soil is exposed to erosion. The thick cover prevents this type of soil erosion in pastures.

Heaving is particularly serious on heavy soils or on those with an impervious subsoil. These soils should be rolled as early

[1] Johnstone-Wallace, D. B., Soil, Field-crop, and Pasture Management for Herkimer County, New York, Part II, p. 63, 1934.

[2] *Idem*, pp. 71–74.

in the spring as soil conditions permit in order to push down high spots and press the roots into close contact with the soil. These soils are in special need of soil treatment to produce a strong growth of grass and wild white clover. Moreover, the clover leaves act like tiny umbrellas, breaking the fall of the raindrops and preventing their beating action on the soil during thunderstorm rains. By increasing the growth of grass the

FIG. 103.—Wild white clover reduces heaving. The Kentucky bluegrass on the left shows very distinct heaving owing to the thinness of the sward. On the right is the thick sward of the bluegrass-wild white clover mixture. Under this cover essentially no heaving takes place, and, consequently, far less erosion than on the soil growing bluegrass without wild white clover. (*Courtesy of D. B. Johnstone-Wallace.*)

clover still further protects the soil. In fact, a thrifty mixture of wild white clover and Kentucky blue- and other grass provides an ideal aboveground protection against erosion.

Underground, wild white clover and grass produce a complete mat of roots, which together with the aboveground growth protects the soil against even a considerable concentration of water passing over it.

As the roots die, organic matter which is thoroughly distributed throughout the surface soil remains in it. This organic matter produces in the soil the advantages enumerated on pages 93 and 94, such as increasing porosity and water-holding capacity. As roots decay in the soil, openings are left through which percolation takes place with increased rapidity. In addition to this, the innumerable stems of clover and blades of grass hold back the rain water until much of it soaks into the soil, instead of running off to the larger streams, as it does from poor pasture swards, thereby aggravating flood conditions.

Practically no water was lost from first-year pasture plots on Lordstown stony silt loam on the Federal Soil Erosion Control Station in southern New York in 1935. The Lordstown stony silt loam normally is an absorptive soil, and, having been plowed in the spring, it would be expected to take up water to an unusual extent during that season. The stones in this soil, by keeping the soil loose, aid in the absorption of water.

Heavy pasture soils and those with an impervious subsoil absorb water slowly, and these soils consequently suffer loss of both water and soil by runoff during heavy rains.

Following is the Cornell pasture mixture for 1936:[1]

Seed Mixture	Pounds per Acre
Kentucky bluegrass (*Poa pratensis*)	10
Canada bluegrass (*Poa compressa*)	2
Rough-stalked meadow grass (*Poa trivialis*)	1
Timothy (*Phleum pratense*)	6
Perennial rye grass (*Lolium perenne*)	5
Wild white clover (*Trifolium repens*, var.)	1
Total	25

A few additional illustrations may be helpful in obtaining a good idea of the use of pastures in controlling soil erosion.

Pastures in Missouri.—Helm and Krusekopf make recommendations for four different upland sections in Missouri in three of which soil erosion is a real problem.[2] Their recommendations for soils of low fertility are the ones that will usually be most useful on the washed hill lands under consideration here.

[1] JOHNSTONE-WALLACE, D. B., Soil, Field-crop, Pasture, and Vegetable-crop Management for Delaware County, New York, *Cornell Univ. Agr. Exp. Sta., Bull.* 639, Part II, pp. 67–70, 1936.

[2] HELM, C. A., and H. H. KRUSEKOPF, Establishing Permanent Pastures in Missouri, *Missouri Agr. Ext., Circ.* 314, pp. 2–5, 1934.

Recommended Seeding Mixtures

Ozark Region:

	Pounds per Acre
Timothy (*Phleum pratense*)	5
Orchard grass (*Dactylis glomerata*)	5
Redtop (*Agrostis alba*)	5
Korean clover (*Lespedeza stipulacea* maxim.)	5
Total	20

In this popular publication Korean clover evidently is used for Korean lespedeza.

Southwestern and Northeastern Missouri Prairie Regions:

	Pounds per Acre
Timothy	10
Redtop	5
White clover	2
Korean clover	5
Total	22

Northwestern Missouri region. Old pastures medium to low in fertility:

	Pounds per Acre
Timothy	5
Orchard grass	5
Red clover (*Trifolium pratense*)	3
Korean clover	5
Total	18

Treatment, including lime, if required, phosphorus, or mixed fertilizer, and soil management under varied conditions are given in the publication cited. The point of similarity with seeding mixtures for the Northeast is the inclusion of legumes for all Missouri conditions, both for the purpose of feed production and for the control of soil erosion. By adding Canada bluegrass (*Poa compressa*) or by substituting it for timothy these mixtures may be useful in the middle Atlantic states.

Pastures in Wisconsin.—For Wisconsin, Graber[1] recommends that poor pastures, which usually include the steep eroded areas

[1] Graber, L. F., Renovating Wisconsin Bluegrass Pastures, *Wisconsin Agr. Ext., Circ.* 277, 1936.

under consideration, be disked, fertilized, and seeded with legumes.

This disking is done in early spring while the soil is loose from freezing and well supplied with moisture. Plant-nutrient treatment is given in accord with the results of soil tests. Lime, if needed, was applied the preceding year. If limestone is used, it is mixed with the soil during the disking. The land is harrowed; phosphorus is applied, and potash, too, if needed; the legume sown; and the land rolled. Rolling covers the seed and packs the soil in close contact with the bluegrass roots. Disking the bluegrass checks its growth temporarily and affords the legumes an opportunity to establish themselves. Biennial white sweet clover (*Melilotus alba*) gives best results for the poorer land, particularly in dry seasons. Wisconsin-grown alfalfa seed or Cossack or Grimm was second, and red clover next, in ability to thrive and to produce feed under Wisconsin conditions. Seeding was at the rate of 20 to 25 pounds to the acre. Alsike and white clover made a poor showing in comparison with the deep-rooting, or *dry-weather*, legumes as they are designated by Graber.

The white grub, which is so troublesome in Wisconsin pastures, was well controlled by sweet clover, alfalfa, and red clover and in that order.

Graber reports the renovation of bluegrass pasture which has been relegated to the steeper slopes and which if plowed might increase the hazards of erosion. The sod was disked, treated, and seeded, as already outlined, in 1930. A torrential rain followed three days later. Even though recently seeded grain fields were gullied, no sign of erosion appeared on the renovated pasture hillside. It is notable that in Wisconsin, as elsewhere, dependence for pasture improvement and production and incidentally for soil improvement and erosion control is placed on legumes and grass as a mixture. In Wisconsin major dependence is placed on these desirable, deep-rooting legumes, of which alfalfa is the longest lived.

Pastures in Kansas.—For eastern Kansas the mixture is recommended for permanent pastures[1] as shown on page 200.

Suggestions for establishing pastures, including the application of lime, if needed, for sweet clover, are given for eastern Kansas.

[1] Lessons for Agronomy Project Leaders, *Kansas Agr. Ext. Service*, 1936

As in the other states, so in Kansas, agronomists place heavy reliance on the use of one or more legumes in pasture production.

Seed Mixture	Pounds per Acre
Brome grass (*Bromus inermis*)	8
Orchard grass (*Dactylis glomerata*)	6
Meadow fescue (*Festuca elatior*)	4
Korean lespedeza (*Lespedeza stipulacea* maxim.) and sweet clover (*Melilotus alba*)	4
(Or 4 pounds to the acre of sweet clover may be used instead of lespedeza.)	
Total	22

This discussion has not been intended as a treatise on pasture production and management. It has been given here for the purpose of showing how thrifty pasture sward controls soil washing on distinctly erosive land and at the same time produces large yields of feed. Pasture treatment and management were necessarily mentioned because of their bearing on the production of the kind of sward necessary for erosion control. Fortunately, this is the identical sward which constitutes excellent pasture. The principles of pasture sward production suggested for the large, important wild white clover region have wide application in their bearing on the control of soil erosion by pastures in general, throughout the United States.

Contour Furrows in Pastures.—The possibilities of the use of contour furrows in pastures on medium- to fine-grained soils deserve full consideration. Furrowing in pastures is important from the standpoints of both water conservation and flood control in humid regions. In the drier areas furrows are important for conserving water for plant growth, but often they aid, also, in flood control, particularly in connection with heavy rains.

The contour furrow is useful in permanent pastures which need all of the rainfall of normal seasons. Usually, however, some summer rain in most sections falls so rapidly that the soil cannot absorb it all. Consequently, water is lost as runoff. If the water that is lost could possibly be held where it falls, the grass would use it to good advantage. As runoff, however, such water intensifies flood conditions in the valleys. Since a simple furrow holds back rain water, increases the pasture production, and at the same time equalizes stream flow and thereby reduces floods, the contour furrow appears certain to attain wide popularity in time.

The contour furrow is most useful in areas that have heavy rains of the thunderstorm type and that have soils of fine to medium texture. Even if some difficulty may be encountered in making the dike, or dam part hold water in stony soils, the contour furrow should be given a thorough trial because of the very great promise that it holds.

Stapleton reports the making of contour furrows by turning four furrows with an 8-inch plow.[1] The furrows are spaced 20 to 30 feet apart. He states that furrows 6 inches deep on a 10-per-cent-slope hold 1¾ cubic feet, or 13.09 gallons, of water for each linear foot of furrow. This is equivalent to about an inch of rain on a 10-per-cent-slope with furrows placed at 20-foot horizontal intervals. In Mississippi he found that during a two- to three-day rain from two to four times the water capacity of the furrows soaked into the pasture soil. The grass showed the effect of this absorption by a greener color among the furrows when it was turning brown owing to lack of water on adjacent unfurrowed soil. In that state the cost of contour furrows is fully warranted either by holding water for use of the pasture grasses or by helping in the prevention of floods. Similar furrows in sparsely timbered areas may be equally as useful as in pastures or meadows for the purpose of flood prevention.

The Soil Conservation Demonstration Project at Meridian, Miss., had made contour furrows on 2,000 acres of pasture on 80 farms up to August, 1935. Such rapid adoption by farmers indicates the practicability of the contour furrow in that state.

Logan reports the development of a new implement for making contour furrows in Kansas.[2] This consists of an ordinary two-bottom plow carriage with a U-shaped blade for cutting loose the furrow slice and bent-steel rods for lifting and guiding the furrow slice over to the lower side of the furrow. Another blade cuts off the surface grass on a strip to which the sod is to be transplanted. A roller, slightly concave, which places greater pressure at the edges, is run over the transplanted sod to press it tightly

[1] Stapleton, D. V., Pasture Contouring Achieves Multiple Results, *U. S. Dept. Agr., Soil Conservation*, Vol. 1., No. 1, pp. 7, 8, 1935.

[2] Logan, C. A., A New Machine for Making Contour Furrows, *U. S. Dept. Agr., Soil Conservation*, Vol. 1, No. 9, pp. 14, 15, 1936.

against the soil. This gives a tighter seal for holding water and encourages a quicker recovery of the grass (Fig. 104).

In Kansas the furrows were made 12 inches wide and 4 to 6 inches deep and 16 to 20 feet apart. Such furrows on the Limestone Creek, Kansas, area increased forage production 29 per cent in 1935.

This contouring implement is particularly well adapted to buffalo and other grass sods which are sufficiently stiff or dense

FIG. 104.—The Kansas contour-furrowing machine. This newly developed implement lifts the sod out of the furrow and sets it over on the side. The roller improves the contact between the slice of sod and the stripped soil underneath. (*Courtesy of U. S. Soil Conservation Service, Kansas.*)

to form a continuous ribbon. A two-way contour-furrowing implement based on the plan of the two-way plow is in process of development.

Lindley gives the effects of contour furrows in Colorado.[1] Range lands have slopes of from 1 to 7 per cent; the soil is dusty; and only about 10 per cent of the torrential rains is absorbed by the soil. Approximately nine-tenths of such rains is lost on lands sorely in need of additional moisture. It is small wonder that the people in the valleys are so deeply interested in the results of contour furrowing on the slopes above them.

In that state the furrows 8 inches deep, 10 inches wide at the bottom, and 20 inches at the top, were made at vertical intervals

[1] LINDLEY, J. C., Protecting Colorado's Range Lands, *U. S. Dept. Agr., Soil Conservation*, Vol. 1, No. 9, pp. 12, 13, 1936.

of 2 or 20 feet apart on slopes of 10 per cent. The soil thrown out of the furrow forms the dike which is set back about 6 inches from the lower edge of the furrow by a special arrangement of the moldboard. The grass immediately adjacent to the edge of the furrows continues to grow and in time covers the raw dike soil and the bottom of the furrow as well. Cross checks were placed by hand every 50 feet in the furrows. Furrows were made at the rate of 6 to 8 miles a day (Fig. 105).

FIG. 105.—Contour-furrowing machine in use in Colorado. An extension moldboard on a disk plow moves the furrow-slice soil several inches over from the edge of the furrow. The soil thrown out of the furrow forms a low dike which increases the capacity of the furrow. (*Courtesy of U. S. Soil Conservation Service, Colorado.*)

Fortunately, the efficacy of the furrows was tested in August, 1935. Two rains of 2 inches, one in 45 and the other in 60 minutes, were held by the furrows without runoff. Under those conditions in Colorado such contour furrows are of inestimable value in flood prevention.

Upon examination of the furrowed soil 60 hours after one of the rains, it was found that the water in the furrows had penetrated the soil to a depth of 41 inches, halfway between the furrows to a depth of 24 inches, and on the same soil still unfurrowed to a depth of 8 inches. The grass made much better growth following rains on the furrowed than on the adjacent untreated area.

Contour furrowing native pastures with a two-row lister has proved very satisfactory in Texas. Provision for reestablishing the sod on the freshly turned soil is imperative, particularly in areas subject to even moderate wind erosion.

Provisionally it appears that the Kansas type of contour furrow is highly beneficial on pastures in many sections of the country. It is better adapted to pastures with a tough sod and soil of moderately fine uniform texture than to thin swards on stony soils. The Kansas type of furrow can be used on a wide range in steepness of slope.

In the Northeast much of the permanent pasture land is steep and stony or rocky. Furrows are out of the question with much outcropping rock or with bedrock at varying shallow depths. On medium-stony land of moderate absorptive capacity contour furrows are useful for the retention of water both for the future use of the pasture plants and for the prevention of floods. Every cubic foot of water held on the land where it falls is that much less water in the valley for causing disaster at times of high water.

Owing to the stones in the soil which tend to break the ribbon of sod, the Kansas type of furrow-making plow may not prove entirely successful in the northeast states. Because of its many advantages, however, the contour furrow should be given a thorough trial in that section. Handwork will be required to patch breaks in the dike caused by stones.

Contour furrows may be made in moderately stony soil with an ordinary two-way plow. The first furrow may be shallow, about 3 inches deep; the second may be made approximately 6 inches deep on the return trip with the second plow. The second furrow is purposely made deeper than the first because the deeper furrow on being turned into the shallow one overfills it and thus builds up a relatively high dike. A jointer is suggested for use on land without many stones or with only small ones. Setting the jointer on the plow that is to make the final furrow about 1½ to 2 inches "in" the furrow slice leaves that much sod hanging over the upper edge of the furrow to protect it from erosion. Establishment of grass in the bottom of the furrow is hastened by this overhanging sod.

On slopes of 25 per cent or more, these furrows if used must be made with care in order to build up a sufficiently high dike to give the furrows a water capacity that is really worth while. In this respect, the shallow first furrow is helpful in keeping the soil from rolling too far down on the slope.

Liming, if needed, fertilizing, and seeding to quick-growing grasses are desirable for the purpose of stabilizing the soil thrown up as dike on the lower side of the final furrow. In the Northeast, a seeding similar to the Cornell pasture mixture (page 197) will be useful. Rye grass (*Loliun perenne*) of the true perennial pasture type and rough-stalked meadow grass (*Poa trivialis*) with true wild white clover (*Trifolium repens* var.) are particularly essential in such a mixture in the area climatically suited to them. Rye grass establishes itself in a few weeks under favorable conditions. This stabilization is highly desirable for preventing the washing down of the dike. It must be kept as high as possible, as previously stated, in order to give the furrow high water-holding capacity.

In the northeast section of the country mid-spring is a good time for contour furrowing. At that season the grass seeding develops rapidly and soon protects the raw soil from washing.

Locating the furrows on the true contour is essential on steep slopes. If the furrows have appreciable slope, their purpose is defeated because the concentration of water scours the bottom of the furrow which in time becomes a gully. Contour furrows are bent uphill at both ends in order to prevent the discharge of water, which might cause erosion on the slope below. Level furrows prevent any concentration of water over a slope; in fact, if level furrows overflow, they act in the same manner as a spreader board.

The contour furrow deserves thorough trial in all sloping pasture areas in the country where it is not now in use. It conserves water for crops and by holding water back on the farm helps to prevent floods.

Hendrickson and Baird report the effects of contour strip plowing in pastures at Tyler, Tex.[1] Narrow contour strips are plowed, leaving alternate narrow strips of the native grass. The plowed strips are spaced 10 to 15 feet apart. The back-furrow ridges form numerous long, narrow catchment basins for water and clover and grass seed. Bermuda grass may be planted in the plowed soil and protected with soil covering until spring. The Bermuda grass grows and spreads rapidly and affords

[1] HENDRICKSON, B. H., and R. V. BAIRD, Summary of Principal Results, Soil Erosion Experiments, Tyler, Tex., *U. S. Dept. Agr.*, SCS-EP-3, p. 4, 1935.

excellent erosion protection. According to the authors, the results from this treatment since 1932 show marked improvement in pasture over previous methods.

Woodman suggests that contour furrows be bent uphill across depressions on all gullied slopes and on those with even slightly depressed watercourses.[1] In that case the furrow takes the shape of a broad U across depressions. The furrow becomes a diversion channel which keeps the runoff water out of the gullies. Consequently, in time, the gullies heal up under even moderately favorable conditions and cease to be a menace to the slope on which they are located.

FORESTS

Steep slopes which are now in forest ought to continue in the production of forest products. Such slopes which are now in cultivation or which have been abandoned as unprofitable should be reforested in all areas that were originally timbered. And as the more productive pasture land is improved, the less easily accessible pasture lands may be planted to trees adapted to the soil and to the climate (Figs. 106 and 107).

The contour idea should be applied in tree planting as in other soil-erosion control work. If, as is often done, trees are to be planted in a furrow, it must be placed on the exact contour; otherwise the furrow almost invariably leads to gully formation down the slope. Furrow planting has an advantage from the standpoint of moisture supply for the trees, but it may be a disadvantage in that the seedlings are planted in subsoil, the surface having been turned out by the plow.

Much tree planting has been done up- and downhill during recent years. Planting in that way may lead to unnecessary erosion between the rows. Cattle and other domestic animals are, of course, excluded from forest plantations. Wild animals, however, may make paths; and if wagons or tractors are driven into a planted area, tracks are made up- and downhill. Water collecting in such depressions during heavy rains starts gullying, as it does elsewhere.

On the Cohocton River Soil Conservation demonstration project in southern New York all tree planting (nearly 450,000 seed-

[1] WOODMAN, V. W., 6th *Soil and Water Conservation Conference, Tyler, Tex., Rept., Oklahoma Agr. and Mech. Coll.*, p. 59, 1935.

FIG. 106.—Natural reforestation. In the background may be seen hemlocks spreading out over steep pasture land. This is desirable from the standpoint of control of soil erosion. Note the healed-up gullies in the foreground.

FIG. 107.—Scotch pines 8 years old. These vigorous Scotch pines are growing on a steep erosive slope. At this size they protect the soil completely. In Delaware County, New York.

lings) in the spring of 1935 was done on the contour in staggered rows. Planting is done more easily and more rapidly on the contour than up- and downhill. Each tree checks water movement in a small way and catches and holds debris which in turn holds the water still further and aids the soil in absorbing it. Thus reforestation reduces runoff, soil erosion, and floods, in time. Weeds and brush grow up and protect the soil after a few years. For several years, however, after recently abandoned bare land is planted to trees, erosion continues essentially unchecked.

Under these conditions on distinctly erodible lands, contour furrows, such as are used in pastures (page 200), may aid greatly in holding water. Based on the judgment of the forester, the trees may be planted in or between these furrows. As water is held in the furrows, it eventually evaporates or percolates into the soil and serves as a supply of water for the growing trees and for springs and wells on lower levels. Whatever water is thus held back on the uplands is that much less water that otherwise might flood the valleys below.

Lands recently taken out of cultivation do not grow up quickly to natural protective vegetation. Even if planted during the season directly following an intertilled crop such as cotton, corn, potatoes, or beans, much washing takes place before the trees afford reasonably complete protection. Moreover, severely washed lands do not produce rapid growth of the trees ordinarily planted, and consequently the most erosive lands suffer the loss of much additional soil before the trees can protect it.

Various means of reducing runoff may be employed during this critical period. In most seasons any conservation of water is beneficial to the trees. In the region adapted to lespedeza it may be sown with a light seeding of grain or with the quick-growing rye grass or other grass adapted to the area. The common lespedeza (*Lespedeza striata*), although an annual, reseeds itself and thus produces a continuing cover for the soil. In areas that have the necessary lime in the soil for biennial white sweet clover (*Melilotus alba*), it may be used in the same manner as lespedeza. Like lespedeza, sweet clover persists under favorable conditions by reseeding itself. Here, as elsewhere, a legume-grass mixture produces the best soil protection. For northern areas, unsuited to either lespedeza or sweet clover, yellow trefoil

(*Medicago lupulina*) may be seeded as outlined for the other legumes. These are merely suggestions for practicable means of controlling erosion in new forest plantations.

Some seedbed preparation for these inexpensive legume-grass seedings is desirable, but it need not be costly. These legume seeds are, of course, inoculated before they are sown.

Moreover, such legume-grass mixtures not only hold the valuable fine soil in place for the benefit of the growing trees, but

FIG. 108.—Black locusts control erosion. These black locusts were planted for the purpose of healing a bad gully. Rye was used with the locusts for temporary stabilization while the locusts were becoming established. These locusts were planted two seasons before this picture was taken on May 7, 1936. This information was supplied by J. M. Dean of the project at Spartanburg, S. C., who is shown standing in the foreground, by a locust tree twelve feet high.

the legumes fix nitrogen. And because of a shortage of nitrogen in badly eroded soils, legumes are an important ingredient in such mixtures.

Another point in connection with planting trees on steep eroding lands is the desirability of including legume trees. Objections are raised to the black locust (*Robinia Pseudo-Acacia*) because sometimes it overtops and smothers other species when interplanted. This is a fine tribute to the black locust, which is capable of making good growth on unproductive soils. The black locust, being a legume, obtains its own nitrogen when inoculated (Fig. 108).

Locusts may be planted in contour strips about 100 feet, or somewhat more, in width alternated with other species of deciduous trees and conifers. Chapman has called attention to the beneficial effects of black locusts on other species of trees.[1] Black locust leaves blown among such deciduous trees as catalpa, white ash, tulip poplar, black oak, and chestnut oak materially

FIG. 109.—Effect of black locusts on growth of non legume trees. Black locusts at the left. Non legume trees, catalpa, at the right. Note the decreasing height of the catalpa trees with increasing distance from the locusts. (*Courtesy of A. G. Chapman. Print used by Chapman in Ecological Monographs, Vol.* 5, *No.* 1, *p.* 44, 1935.)

increased their growth both in height and in diameter immediately adjacent to the locusts, as compared with distances of from 12 to 72 feet away (Fig. 109). Precisely the same relationship was found in the nitrogen content of the soil, it being 0.196 among the locusts and 0.090 per cent at a distance of 72 feet from the locusts. As Chapman points out, this difference may appear small, but it is equivalent to 3,900 pounds of nitrogen to the acre among the locusts and to 1,800 pounds an acre in the surface 6 inches of soil (weighing approximately 2,000,000 pounds over an acre). In other words, the locusts appear to have increased the nitrogen content of the soil by 2,100 pounds to the acre or to

[1] CHAPMAN, ARTHUR GLENN, The Effect of Black Locust on Associated Species with Special Reference to Forest Trees, *Amer. Soil Survey Assoc., Bull.* 15, pp. 39–41, 1934.

have more than doubled it. This difference in nitrogen content of the soil among the locusts as compared with that 72 feet away fully explains the additional growth of nonlegume trees which were planted near the locusts.

Chapman gives the nitrogen content of elm leaves in Ohio as 2.06 per cent and of a mixture of elm and locust leaves on Oct. 12 as 2.46 per cent. Black locust leaves collected in Ithaca, N. Y., Oct. 9, 1934, contained 2.33 per cent of nitrogen on the basis of dry matter,[1] which is essentially identical with the nitrogen content of first-cutting red clover at full bloom.

Ebermayer[2] reports that mixed beech and pine in France produced 2,800 pounds of dry matter to the acre in leaves and needles and that beech produced 3,000 pounds of dry leaves to the acre. If the black locust produces approximately this quantity of leaves, they contain about 70 pounds of nitrogen to the acre, or, on the basis of only a ton of leaves, 46.6 pounds to the acre. This important contribution of nitrogen and organic matter to the soil is highly beneficial to nonlegume species.

When locust leaves are blown over on to adjacent strips of other trees, the addition of nitrogen thus made available aids in rapidly reclothing eroding lands with the perfect protection that the original forest trees provided.

Locusts are needed in limited numbers to provide posts for farm fencing and for such use as telephone poles. The locusts, if not needed or desired for these or other purposes, may be held down by cutting, after they have served their main purpose of aiding in quickly bringing erosion under control.

On badly eroded areas, regardless of the slope, the solid planting of black locusts is warranted. Some soil protection against erosion by means of a light covering of straw, pine needles, cane pomace, or other decay-resisting material is desirable. Under favorable conditions, however, closely planted locusts produce soil protection in a few years. Once erosion is under control, the locusts may be cut down, and other species planted if other trees are more valuable or more desirable.

[1] GUSTAFSON, A. F., Composition of Black Locust Leaf Mold and Leaves and Some Observations on the Effects of the Black Locust, *Jour. Amer. Soc. Agron.*, Vol. 27, pp. 237–239, 1935.

[2] EBERMAYER, ERNEST, Naturgesetzlich Grundlage d. Wald-und Akerbaues, *Physiol. Chem. der Pflanz.*, Vol. 1, pp. 41–44, 1882.

CHAPTER XIII

CONTROL AND ELIMINATION OF GULLIES

Gullies are formed in sloping cultivated fields by the concentration of water often as the result of accidents or of errors in management. During the progress of tillage operations, however, it is not possible to be always prepared for torrential rains. Extremely large quantities of water from heavy rains, pouring down over slopes, especially if carrying sharp, angular stones, cut gullies in meadows and pastures. Thin swards suffer more than do thick thrifty ones. Gullying generally follows the loss of surface soil through sheet erosion.

Gully prevention, which has already been discussed, is far better than gully control; but since gullies do form, control measures must be considered.

On the average good farm, if gully control measures are to be put into effect, such measures must be simple, effective, practical, and withal inexpensive. Moreover, they must not require any considerable time or expenditure for maintenance. Land subject to severe gullying is usually not highly productive and is more or less difficult to till. Consequently, gullied land cannot pay for expensive gully-control works or for any considerable maintenance of them. Vegetative control of gullies, therefore, is of utmost importance from the farmer's standpoint and for that reason receives major consideration on the succeeding pages. Numerous situations arise, however, where engineering structures are useful, if, indeed, not indispensable. Such situations are discussed particularly in connection with public works such as reservoirs, highways, and streams.

The general principles of erosion control apply in the management of gullies the same as elsewhere, namely, in maintaining a continuous blanket of vegetation between the soil and the water flowing over it, in reducing the quantity and velocity of water in the gullies by means of diversion ditches, and in paving or erecting control structures if neither of the preceding measures can be made effective. Many phases and variations of these general principles are employed in the control of gullies. The necessity for preventive measures, however, must be kept ever in mind in farming rolling lands.

Causes of Gully Formation. *Wheel Tracks.*—Wheel tracks made up and down or angling down slopes are a suitable place for the concentration of water and the starting of gullies. Tracks

FIG. 110.—Up-and-down hill wheel tracks. These wheel-track gullies are not serious yet, but might readily become so with heavy rains. Photograph taken in central Illinois.

angling down a slope, particularly if up-and-down crop rows cross the wheel tracks, concentrate much water in the upper track. Such a situation is not uncommon and is particularly serious because it leads to severe gullying by heavy rains (Fig. 110). During harvesting operations farmers often lock one wagon wheel while hauling the crop downhill. The locked wheel makes a small trench, the bottom of which is compacted so that the soil does not take up water readily; and such tracks angling down the slope, as just mentioned, often lead to the formation of a sizable gully during a single rain. The solution is the avoidance of the formation of up-and-down hill tracks. Harvesting, like all other field work, is best done on the contour in so far as possible and feasible.

Cattle Paths.—Cattle in going to and from a hill pasture follow the same unobstructed way until a smooth sloping path has been formed. A path angling across a slope intercepts water coming directly down the hill exactly as do wheel tracks, as indicated in the previous paragraph. The water is concentrated in the path toward the base of the hill to such an extent as to give the water high velocity and tremendous cutting and carrying power. Gullies are the result.

To prevent such gully formation, obstructions such as brush and many sorts of otherwise useless materials may be placed in the paths. Placing straw or fine brush in the paths in connection with obstructions reduces erosion.

Another plan, where the formation and use of paths cannot be avoided as in narrow up-and-down hill lanes, is to cut the edge of the path at short intervals so as to drain the water from the paths on to adjacent pasture sward or other stabilized soil. Thus the concentration of sufficient water to cause serious washing is prevented. Widening lanes or relocating them on gentler slopes often helps prevent cattle-path erosion.

Fertilizing pasture lanes with an abundance of phosphorus, in addition to the cattle droppings, strengthens the sward, especially if a legume is present, to such an extent that erosion is reduced or is sometimes prevented.

Dead-furrow and Other Up-and-down Hill Depressions.—Dead furrows up- and downhill lead to the collection of water and the formation of gullies. Any open furrow or depression made by planters, harrows, cultivators, or other implements concentrates water and may result in washing. Depressions that angle down a slope present an especially difficult situation in that much water is concentrated in such places. Severe gully formation often follows such concentration of water.

Diversion Ditches.—Diversion ditches or diversion terraces are used to keep the water out of gullies and to spread it on well-protected soil. Sufficient length of diversion ditches is essential so that the water may not find its way back into the actively eroding gully a short distance below the point of diversion. Outlets for this type of diversion ditch require attention in order to avoid the formation of new gullies. The water should be spread to deprive it of cutting power. The quantity of water intercepted determines the type of outlet protection that is needed. Gullies formed by the occasional extremely heavy thunderstorm rain heal up in time by natural means if the cropping system is changed so as to avoid the concentration of water in them. The beneficial effects of contour tillage and planting

FIG. 111.—Diversion ditch to stop gullying. This ditch diverts the water which collects on the slope above and conducts it to a safe place of disposal in the forested area in the background. In North Carolina, lespedeza seeded under a light protection of straw or pine needles soon reclothes an area such as this with protecting vegetation. (*Courtesy of U. S. Soil Conservation Service, North Carolina.*)

in preventing the concentration of water at any point on slopes cannot be overemphasized (Fig. 111, also 78 and 79).

USE OF VEGETATION IN GULLY CONTROL

In addition to living, growing vegetation, straw and brush, although not in a vegetative state, are included under this heading.

Straw.—The use of straw in small- to moderate-sized gullies is an old practice which is still as good as ever (Figs. 112 and 113). The straw is trampled into small gullies; but if nothing further is done, the first heavy rain may wash it out again. To avoid this, stakes leaning uphill are driven through the straw into the bottom of the ditch. The weight of the water presses the straw tightly against the stakes instead of floating it out, as might occur in the absence of stakes. Stakes set so as to hold the straw against the sides of the gully are especially helpful. In a gully as much as 2 feet deep it is well to drive the stakes into the bottom

FIG. 112.—Straw checking gullying in southern Illinois. A gully such as this should be seeded to a suitable grass-legume mixture and left as a permanent grass waterway. Photographed, 1912.

FIG. 113.—Another straw-filled gully, in western Illinois. Stakes are needed to hold the straw in this gully. It may be plowed in on the straw and seeded, to be maintained as a permanent grass waterway. This gully must have immediate attention, else the corner of the field will be cut off, which will increase costs of crop production.

of the gully to a depth of 1½ or preferably 2 feet. After prolonged rains the soil in the gully becomes wet and soft. Because the pressure of the water on the straw against the stakes is relatively great, the stakes may be pushed over unless they are driven into the soil to a sufficient depth. If this occurs, the straw is washed away along with any soil that it may have collected by filtering it out of the runoff water.

FIG. 114.—Brush in gullies in western North Carolina. Brush has been placed in these gullies and is controlling them successfully.

Brush.—Brush in addition to straw is used in somewhat larger gullies. The straw is placed on the bottom of the gully with brush on the straw. Rocks or logs are placed on the brush to hold it and the straw in place. On relatively steep slopes, stakes, as used in straw, are needed to hold the straw-brush fill from being washed away (Figs. 114 and 115).

Live or growing stakes have many advantages over ordinary ones. Cottonwood (*Populus deltoides virginiana* (*Castigl.*) *Sudw.*) is used in the South in the latitude of Mississippi and (*Populus balsamifera*) in the North. Cottonwood cuttings from the present or previous year's growth may be planted in the brush toward the edges of the gully. These cuttings take root under conditions of sufficient moisture and grow rapidly. A sufficient number may be used to form a sort of growing dam across or a living dike along the gully on pasture or on other uncultivated land. Once they are well started, cottonwood cuttings grow on rather unfavorable situations such as on dry sandy soil. An adapted willow (*Salix* spp.) may be used in much the same way but in somewhat wetter soil (Fig. 116). Large willow stakes or posts may be used in the middle of the gully for holding brush and straw in place. Some of the willow stakes take root and

FIG. 115.—Gullies in eastern Missouri. Some brush and trash in the left fork of this gully are checking erosion. The main gully is susceptible of control with brush.

grow and hold the soil long after the straw and brush have decayed. Such living stakes do not require replacing and become more efficient as they grow. If the top growth becomes troublesome in any way, it may be cut off. New growth comes from the root or stump and continues the control of erosion. On the drier, upper part of gully slopes, black locust (*Robinia Pseudo-acacia*) is remarkably helpful in gully elimination. It grows rapidly and encourages other stabilizing vegetation by means of the nitrogen fixed by the locust and spread over the soil in the locust leaves.

Grasses and Legumes.—The seeding of adapted grasses and legumes alongside of the brush is advantageous in forming a grassed limit on the width of the gully. This grass prevents the formation of a new gully parallel to the old stabilized one.

In places, it is best to fill a small gully with straw and then plow it shut. If no straw can be spared from other uses, the gully is plowed in and seeded to a mixture of adapted legumes and grasses. When these become established, the gully bottom is left in sod. In other words, it becomes a grassed watercourse and should not be plowed out again for many years. As pointed out elsewhere, the mixing of legumes with grasses makes the sward longer lived or more nearly permanent. Unless the adjacent farm land is well

FIG. 116.—Willows check waterfall erosion. Note depth of gully at left. The willows prevent further extension of the gully. Photographed, 1914.

fertilized, the application of phosphorus to this grass helps in the development of a thick erosion-resisting sward. The grass as stated under *Grassed Waterways* (page 145) catches and holds much soil by filtering it out of the water which passes down the grass-covered channel. After gullies, which are filled with straw and brush, become partly filled with soil they also may be seeded and managed in this manner under favorable conditions (Fig. 117).

Bermuda grass (*Cynodon dactylon*) and centipede grass (*Eremochloa ophiuroides*) are highly effective in checking soil erosion in the Gulf states. Bermuda grass spreads by means of both underground rootstocks, or rhizomes, and creeping stems, or stolons; while centipede grass spreads by means of surface runners, or stolons, which root at the nodes. Farmers in the South have objected to Bermuda grass much as farmers in the North have to quack grass (*Agropyron repens*), owing to its rapid-spreading, underground growth habits. This objectionable feature of these two grasses in cultivated fields is the property that makes them so highly successful in controlling and eliminating erosion. The planting of stolons or sod of Bermuda and

FIG. 117.—Sweet clover and grass check erosion in southern Michigan. Seeding sweet clover (*Melilotus alba*) and Kentucky bluegrass (*Poa pratensis*) in gullies of this nature hastens their stabilization.

centipede grass gives the quickest control of erosion, but Bermuda grass may be established by seeding. Centipede grass produces but little seed (Figs. 118, 119, and 120).

Winters[1] reports that "sedge" (*Andropogon* sp.) growing in a gully of U-type cross section, draining 20 acres on a slope of 9 per cent, has produced silt deposition in Missouri. Surely that grass has rare possibilities for gully-control purposes in that section.

Native wild legumes may be encouraged to special advantage in the southern states. Certain legumes seeded with the grasses produce the same desirable results in the South as others do in the North. Among them are the lespedezas, especially the com-

[1] See reference on p. 165.

Fig. 118.—A man-induced gully in North Carolina. Erosion is very active in many gullies of this nature. (*Courtesy of U. S. Soil Conservation Service, North Carolina.*)

Fig. 119.—Gully treatment in North Carolina. I. The gully shown in Fig. 118 was seeded and a straw mulch placed on the steep sides. Note the use of straw and brush held by posts in the bottom of the gully. Locusts and pines are planted on the banks. (*Courtesy of U. S. Soil Conservation Service, North Carolina.*)

mon lespedeza (*Lespedeza striata*), and the piedmont butterfly pea (*Bradburya virginiana*), the porcelain butterfly pea (*Clitoria mariana*), the partridge pea (*Cassia fasiculata*), *Crotolaria* sp., tick clovers (*Meibomia* spp.), and vetches (*Vicia* spp.). The common lespedeza grows wild in the latitude of southern Illinois, and the partridge pea north to that of the central part of that state. Although these legumes may not be especially useful on

FIG. 120.—Gully treatment in North Carolina. II. Comparative treatment of a vegetative nature is in progress. Locusts, pines, and lespedeza have been planted and the surface mulched with pine forest litter. These gullies must be stopped before they ruin more good land.

steep gully banks, they grow well on land whose surface soil has been largely lost and on which gullies are beginning to form. They can aid nonlegumes in getting a firmer foothold under poor soil conditions.

Sod Bags.—Uhland has described the use of sod in burlap bags for the control of small gullies.[1] Old bags with holes are suitable for this purpose. In the Midwest the bags are partly filled with bluegrass sod. Bermuda-grass sod may be used in the South if the farmer does not object to it too strenuously. The bags are

[1] UHLAND, R. E., Controlling Small Gullies by Bluegrass Sod, *U. S. Dept. Agr., Leaflet* 82, 1931.

laid across small gullies with the ends up on the sides to prevent the water from cutting around them.

Sod bags should be particularly helpful in stopping headwater-erosion in small newly formed gullies in grain and in cultivated fields. The bag of sod is tramped against the waterfall tightly so that water does not pass between it and the soil. The grass grows out through the bag and soon clothes the waterfall with grass. Thus erosion is stopped at least for the time being. This method of controlling erosion appeals to farmers because it is effective and because it entails no cash outlay.

FIG. 121.—Waterfall erosion in southern Illinois. These gullies menace the farm buildings. The drainage area is small. This picture was taken by the writer just over the fence from the state soil erosion-control field in 1911.

Control of Waterfall Erosion.—Waterfall, overfall, and gully-head erosion refer to the same waterfall type of gully extension back up into a slope. Once started, gullies grow rapidly in this manner, particularly if the underlying material is nonresistant to erosion. High check dams, brush dams, and diversion ditches are useful in combating this phase of erosion.

The method shown in Figs. 121 and 122 is an adaptation of methods in use. The old perpendicular fall is cut down so as to establish the new surface along the line AC which has a fall of from 35 to 40 per cent. The sod or turf is laid aside. The pool cut out by the waterfall is filled with soil from above the fall. In small gullies the work may all be done with a spade and shovel. In larger ones a plow and slip scraper may be used advantageously. After the face of the fall has been sloped, the turf may be planted in bands across the slope or irregularly over the area. If such work is done in the spring, spring grain and a suitable adapted grass-legume mixture are seeded over the exposed soil (see pages 146 and 149); if in midsummer to late summer, rye and the grass-legume mixture are used; if done in early fall, a thick seeding of rye may be all that is feasible until spring when the regular grass-legume mixture is sown. Throughout the entire area to which it is adapted a rye grass is recommended, because

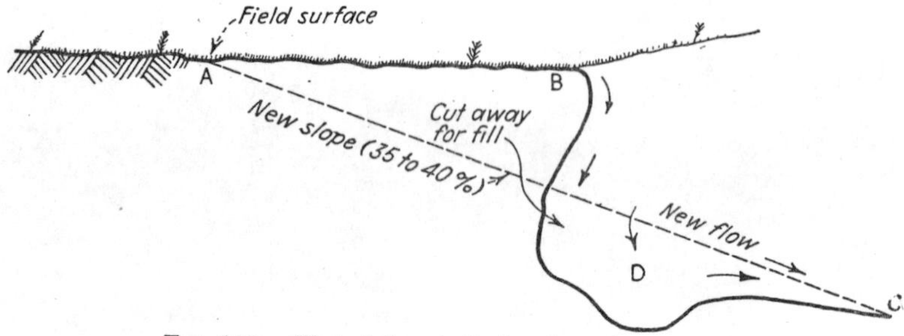

FIG. 122.—Waterfall or gully-head erosion control.

of the very rapid growth that it makes. In its area rye grass provides protection for the soil in an unusually brief period (Figs. 123 and 124).

A light coat of manure or straw over the seeded area helps in holding the soil until the grain, legumes, and grasses make the necessary growth for erosion control. Fertilization, in addition, thickens the sward and thus improves its control of erosion. Once such situations are stabilized, they should remain in permanent sod.

Vines.—In many situations vines can be used to better advantage than shrubs. They may be interplanted with shrubs and thus control erosion more quickly and more completely than do shrubs planted alone. The Hall Japanese honeysuckle (*Lonicera japonica Halliana*), wild grapes (*Vitis* spp.), and the Virginia creeper (*Ampelopsis quinquefolia*) grow over an especially wide range of latitude. The aboveground growth of the Japanese honeysuckle winterkills in northern states in severe winters, but this plant produces a luxuriant growth from the root as far north

FIG. 123.—Gully head in North Carolina. Dynamite was used for sloping the sides of this large gully. Holes for the charges are being dug. (*Courtesy of U. S. Soil Conservation Service, North Carolina.*)

FIG. 124.—Completed gully sloping. The gully shown in Fig. 123 after sloping treatment. The slope on the sides was reduced to 3 to 1, or 33 per cent, March 11, 1936. This gully is ready for seeding and mulching. Lespedeza and locusts could be very helpful in a situation such as this. (*Courtesy of U. S. Soil Conservation Service, North Carolina.*)

as central New York. Farmers raise objections to it in parts of the South because of its encroachment on cultivated lands. Its rapid growth, however, makes this plant exceedingly useful in gully control over large areas. Wild grapes, the summer grape (*Vitis aestivalis*), fox grape (*V. labrusca*), and frost grape (*V. vulpina*), American bittersweet (*Celastrus scandens*), and carrion flower (*Smilax herbacea*) are hardy and because of their rapid growth quickly cover gully banks. The Virginia creeper is useful under various conditions on gullied and other badly washed lands. The wild dewberry (*Rubus* spp.) grows moderately well on badly washed soils where gully formation is underway (Fig. 179).

Kudzu.—Kudzu (*Pueraria thumbergia*) is a most promising soil-binding gully-control plant for the South to which it is limited. Kudzu is a legume, introduced from China, which when established on unproductive eroded soils makes a growth of 20 to 50 feet in a single season. The vines root deeply at the nodes and cover the soil, climbing over grass, shrubs, fences, and into trees. It fixes much nitrogen for other plants. In open fields it furnishes much palatable forage for livestock. In the bottom of gullies sand or soil is washed over the vines, but they come up through the deposit and continue their growth. Its ability to control gullies is indicated by a representative situation related by Meginnis.[1] Kudzu roots were set in a gully on a farm near Canton, Miss., in 1926. In four years the vines had spread to adjoining gullies and had covered several acres with a dense growth. Active erosion had been stopped. Several gullies had been filled with soil 10 feet deep as the vines caught and held the soil washed from the gully banks and adjacent land (Fig. 125).

Shrubs.—Planting shrubs, alone or in combination with grasses and legumes, and trees under some conditions, on the sides of gullies aids in the control of erosion. Native or indigenous shrubs or introduced ones which have escaped from cultivation and are now growing in the wild are available in many sections. Most wild shrubs are indigenous. Shrubs, which may be obtained on the farm or from fence rows or roadsides without cash outlay, may be used freely. If transplanted at the proper time, these shrubs from near-by areas continue to grow

[1] MEGINNIS, H. G., Using Soil-binding Plants to Control Gullies in the South, *U. S. Dept. Agr., Farmers' Bull.* 1697, p. 11, 1933.

almost as if they had not been disturbed. Any shrub that has a spreading habit in its native locations may be expected to grow and spread in gully-control plantations (Fig. 89).

If gully-control values are approximately equal, those shrubs which produce human food or feed for birds or other animals and which provide shelter for desirable wild life deserve preferred consideration.

FIG. 125.—Kudzu controls gullies in South Carolina. Twelve kudzu plants were set around the sides of this gully in 1922. Six plants survived. The gully was originally 35 feet deep and 30 feet wide. It is now 20 feet deep and is filling up at the rate of about 3 feet a year. A wide extension of use of kudzu in gullies and in other eroded places in the South appears fully warranted. Photographed on farm of H. H. Arnold, four miles west of Woodruff on the South Tyger, U. S. Erosion Control Project. (*Courtesy of U. S. Soil Conservation Service, South Carolina.*)

Useful Shrubs in the Northern States.—Representative of the first group adapted to the northern states may be mentioned the common blackberry (*Rubus allegheniensis*), the black raspberry (*Rubus occidentalis*) and the red raspberry (*Rubus idaeus*), the black huckleberry (*Gaylussacia baccata*), and the high-bush blueberry (*Vaccinium corymbosum*). All of these make good growth under adverse conditions, produce edible fruit, and furnish excellent shelter for wild life. The common elder and the red-berried elder (*Sambucus canadensis* and *S. racemosa*), respectively, grow well in places somewhat wet in the spring.

The wild plum (*Prunus americana*) may be used in much the same manner as these adapted wild berries. Some of the wild plums are very useful, particularly in sections where few fruits thrive.

The hazelnut (*Corylus americana*), while functioning somewhat differently, produces human food and squirrel feed and wild-life shelter. Some hazel hybrids produce more desirable nuts than do the wild ones.

Some of the dogwoods grow and spread on raw banks and hold the soil against washing. Among the useful ones are the gray, silky, and red-osier dogwoods (*Cornus candidissima*, *C. amomum*, and *C. stolonifera*), respectively, but in addition to gully control this group provides only shelter for wild life. The dwarf bush honeysuckle (*Diervilla lonicera*), the coralberry or the buckberry (*Symphocarpos orbiculatus* and *S. vulgaris*), and numerous other shrubs are probably just as useful as the dogwoods. The redbud or Judas tree (*Cercis canadensis*) is too large for many situations, but it is a legume and for that reason deserves consideration. The black locust (*Robinia Pseudo-acacia*), also, is large and, owing to its unusually rapid growth on raw subsoils, when it is thoroughly inoculated, can be planted to good advantage for purposes of gully control. It may be held to shrub size by cutting it back as necessary, wherever locust trees are objectionable.

Useful Shrubs in the Southern States.—Many of the foregoing shrubs grow over a considerable north and south range. The persimmon (*Diospyros virginiana L.*), sassafras (*Sassafras variifolium* (Salisb.) Ktze.), and Chickasaw plum (*Prunus angustifolia Marsh.*) grow well on poor eroded soil. The sassafras and the plum spread by root suckers to form thickets. The persimmon, where the writer has observed it, is more inclined to produce trees than thickets. Its ability to grow on eroded land, however, gives it real value for purposes of reclamation of such lands.

The sassafras and persimmon may, if permitted, grow entirely too large for ordinary-sized gullies. If planted thickly, however, the former grows mainly as a tall shrub. But, as suggested for the black locust, this and persimmon may be held to the maximum size desired by cutting. The sassafras grows as a shrub or a small tree and sometimes attains the size of a large tree.

The habitat of these species overlaps that of the shrubs listed for the North to a large extent. Sassafras is found as far north as central Illinois and occasionally in upstate New York. The persimmon is common in south-central Illinois but has not been observed by the writer farther to the northward.

Trees.—Trees may be used to advantage mainly in very large gullies, in gullies in pasture, or in those which for unusual reasons occur in wooded areas. Species to be used are best confined in most places to those which are thoroughly acclimated to the area in which they are to be planted. As already indicated, legume trees, black locusts, and, to some extent, redbud have a place

FIG. 126.—Darby-High gully treatment in South Carolina. This gully had cut back and caused the moving of the public highway to its third location. The sides were cut down, diversion ditches made for conducting the road water into wooded area at the right. Bands of Bermuda-grass sod were placed in the ditch. Small grain was seeded and black locusts were planted 4 feet apart. (*Courtesy of U. S. Soil Conservation Service, Spartanburg, South Carolina.*)

for gully planting either solid or mixed with nonlegume trees. If trees are not desired after they have controlled gully erosion, they may be removed, provided, of course, that sufficient vegetation is left to keep the gully banks stabilized (Figs. 126 and 127).

The Use of Conifers.—Some of the indigenous conifers may be used to advantage. While some of them have the reputation of not surviving transplanting well, the hemlock (*Tsuga canadensis* and *caroliniana*) in its habitat grows well after being transplanted. In the South the shortleaf pine (*Pinus echinata* Mill.) grows on poor soils and is useful on badly gullied lands. More care is required in transplanting it than the locust. Alternate planting

FIG. 127.—Locusts in Darby-High gully. Two years of growth on locusts. Rye was seeded in fall of 1934. Photographed by writer with J. M. Dean, May 7, 1936.

FIG. 128.—Conifers on badly eroded area in southern Ohio. Here sheet erosion was followed by severe gullying. Stone and log check dams have been well placed. Pines have been planted. Erosion continues. Seeding a legume and grasses, or small grain, or all three, and the planting of locusts are needed to complete the control of erosion on this slope.

of the shortleaf pines with black locusts is suggested by Meginnis[1] (Fig. 128).

If the locusts threaten to overtop and choke out the pines, the former may be cut down, because they sprout profusely from the stumps, and in addition they spread by means of suckers. Thus they continue to supply nitrogen to the pines and other non-legumes as well as to hold the soil against loss.

DAMS

Dams of various kinds, types, and materials are useful in connection with vegetation in controlling and filling gullies. Under conditions of severe erosion the erection of one or more dams may be the first and the most essential step in gully control. In other situations temporary structures may be of service, mainly, in aiding vegetation to bring about permanent stabilization.

Brush Dams.—Brush dams are a variation of the combined use of straw and brush previously advised for small gullies. Brush dams for checking the flow of water and causing silting have been somewhat generally used successfully in gullies up to 4 or 5 feet in depth and even 7 feet deep in Nebraska, according to Wood.[2] The use of straw and brush fills gullies most rapidly in watersheds that are planted partly to intertilled crops. It is the area of the latter that supplies the soil material for filling the gullies (Figs. 129 and 130).

A distinct advantage is that brush dams are inexpensive. Brush, posts, a little wire, and a few staples are all that is needed for making these dams. Essentially no cash outlay, therefore, is required of the landowner.

Wood gives the following list of materials needed for making a brush dam with a single row of posts in a gully 3 feet deep and 10 feet wide:

3 sound posts with 4-inch tops, 6 feet long
4 stakes, 3 inches in diameter, 3 feet long
2 poles, 3 to 4 inches in diameter, 6 feet long
20 staples
15 feet of No. 9 wire

[1] *Ibid.*, p. 7.

[2] WOOD, IVAN D., Inexpensive Methods of Gully Control, *Nebraska Agr. Ext.*, *Circ.* 741, p. 6, 1933.

1 hay rack load of brush (preferably freshly cut)
1 double wagon box full of . . . wet straw. . . .

A dam built of these materials is indeed inexpensive on a farm having a woodlot that can furnish the necessary brush and posts.

Constructing the One-row Post Brush Dam.—The three posts are set to a depth 3 to 3½ feet to give strength for holding the brush against the pressure of the water. The sides of the gully are sloped to an angle of 45 degrees, or 100 per cent, or somewhat

FIG. 129.—Gully treatment with brush. I. Note the sediment caught and held by the brush. (*Courtesy of U. S. Soil Conservation Service, North Carolina.*)

less at the upper edge. The four stakes are next set to a depth of about 2 feet, two on each side of the gully in line with the posts (Fig. 131). The wet straw is placed on the bottom and well up over the top of the sides in order to keep the water from flowing under or around the brush. The longest brush or branches are laid first in the bottom of the gully, the forks being hooked over the posts. The butt end of the brush extends upstream a foot or two from the posts. The important point is the fork hooked over the posts for anchoring the brush at as many places as possible in the dam. The longer branches, in fact, constitute an apron for protecting the bottom of the gully below the completed dam. The brush is built up on the sides of the gully with

FIG. 130.—Gully treatment with brush. II. These brush dams have caused silting and have prevented further erosion. This gully is almost ready for seeding to a mixture of legumes and grasses. (*Courtesy of U. S. Soil Conservation Service.*)

FIG. 131.—Building a brush dam. I. The posts are in place for the one-row post brush dam. (*Courtesy of Ivan D. Wood. Nebraska Agr. Ext. Service.*)

a distinct depression left in the middle for the passage of the water. All of the water must be kept on the brush in order to prevent any cutting around the dam. One man works on the brush, packing it at all times, and this is the reason for using fresh rather than old, brittle brush. Close packing is absolutely essential. If the brush is too coarse or stiff to admit of close packing, a little dry rye, or other straw, leaves, or cane pomace may be mixed with the brush to improve its effectiveness as a filter. With the weight of two men on them, one after the other of the 6-foot poles is secured in place with the No. 9 wire provided

FIG. 132.—Building a brush dam. II. The brush has been placed and wired down under the poles. (*Courtesy of Ivan D. Wood, Nebraska Agr. Ext. Service.*)

(Fig. 132). Wet straw is worked into the butt ends of the brush on the upstream side of the dam to catch the silt and seal the dam (Fig. 133), which then is ready for use.

The soil removed from the upper edge of the gully sides, while easing the slopes before placing the wet straw, may be rolled into the gully. The sod or turf removed may be planted at the lower edge of the brush apron. This is particularly desirable if there is not already a stiff erosion-resistant sward in the bottom of the gully.

Constructing the Two-row Post Brush Dam.—The double-row post brush dam is used in gullies 7 to 8 feet in depth and 20 feet or more in width which, according to Wood, may have a watershed of 100 acres or more. The dam and the apron are made very much as is the single-row post brush dam. Owing to the great strain, the middle posts are set to a depth of 4 feet or more in a large gully. The two rows of posts are set from 2 to 2½ feet apart, and the posts are spaced from 1½ to 2 feet apart in the row. The finer brush is tramped in between the two rows of posts and stakes on the butt ends of the brush, which forms the apron and the lower part of the dam. A wire stapled to an end stake is wound back and forth and stapled to the posts. While stapling, the brush is held down by the weight of men as with the smaller dam. The cross brush of the dam is thus held securely

FIG. 133.—Brush dam completed. III. The dam has been completed and the wet straw placed for catching the silt washed from the fields above. (*Courtesy of Ivan D. Wood, Nebraska Agr. Ext. Service.*)

in place by means of these wires. Figure 134 shows a two-row post brush dam in place after a heavy rain.

Because the apron is long in order to serve as ample spillway protection, a row of stakes across the gully is driven into the bottom of the gully below the double row of posts. A heavy wire is fastened to one end post. As one man holds the brush down by standing on it, another stretches the wire from post to post and staples it to each in turn. This wire holds the brush apron in place.

In gullies that have a bottom slope of about 3 per cent, the brush dams are placed approximately 100 feet apart. Silting takes place above one dam and extends from it to the dam next above, especially after vegetation has obtained a foothold in the sediment. Some gullies have been filled with soil during a single rain. Such rapid silting, however, indicates the presence of highly undesirable erosion conditions on the watershed.

Willow stakes may be used in the apron. Such stakes take root and grow if, as often happens, sufficient moisture is present much of the time. Early in the spring, willow and cottonwood cuttings may be planted in brush dams of this sort. As the

FIG. 134.—Effect of a brush dam. Here is shown the effect of a brush dam in catching sediment and in controlling erosion. This two-row post dam is used in larger gullies than is the one-row post brush dam. Such dams have proved to be very effective wherever used. (*Courtesy of Ivan D. Wood, Nebraska Agr. Ext. Service.*)

cuttings grow up through the brush, they produce a mass of roots and a thick top growth. In fact, this growth constitutes a living dam to replace the brush dam when the brush decays. Such living dams control erosion for a long time. Except, perhaps, where a grassed watercourse is capable of controlling erosion in a field growing intertilled crops, after gullies have been filled, such vegetative dams should not be disturbed. If such land is plowed and planted to clean-cultivated crops, there is constant danger of recurrent gully formation.

Earth Dams.—Earth dams are built in gullies or ravines to catch and hold soil and to control gullies. The term *soil saving* is often applied to these dams. Earth dams may be built from a few to 40 or 50 feet or more in height. Only the small to medium ones are considered here. If the drainage area or watershed is more than from 10 to 15 acres, the entire situation should be studied, and the plans worked out by a competent experienced agricultural engineer. Moreover, the cost of large dams is usually too great to be borne by an individual farm of average size.

The first step in the building of a dam for this purpose is to determine with a fair degree of accuracy the acreage of the watershed. Table 31 shows the size of tile required to let the water

TABLE 31.—REQUIRED SIZES OF DROP-INLET CULVERTS FOR SOIL-SAVING DAMS*
(Diameter in inches.)

Watershed area, acres	Rolling land		Hilly land	
	5-foot drop	10-foot drop	5-foot drop	10-foot-drop
	Cultivated land			
2	12	12	15	12
4	18	15	18	15
6	24	18	24	18
8	24	18	24	24
10	24	24	30	24
	Pasture land			
2	. .	. .	12	
4	12	12	15	12
6	15	15	18	15
8	18	15	18	18
10	18	18	24	18
	Timber land			
2				
4				
6	12	12	12	12
8	12	12	15	12
10	15	12	15	15

* RAMSER, C. E., Gullies; How to Control and Reclaim Them, *U. S. Dept. Agr., Farmers Bull.* 1234, p. 32, 1935.

out of the temporary reservoir with sufficient speed for safety under most conditions.

The acreage of the watershed, its slope, and its land cover are all taken into consideration in determining the required size of the outlet. It is better to use tile that is somewhat larger than is absolutely necessary rather than tile of inadequate capacity. For the size of dam under consideration, No. 2 vitrified tile is regarded as satisfactory, the cost of No. 2 tile being materially lower than that of No. 1. All trash, weeds, grass, debris, wood, sod, or other material, which might aid the water in finding a way to get under the dam, should next be removed. For the larger dams of this size range, plowing a deep furrow across the gully where the middle of the dam is to be located helps in producing the desired seal at the base of the dam. In the smaller dams and those which may be expected to fill rapidly this precaution is less essential. The removal of debris, however, is always advisable.

A groove is made for the tile, with a special cross slot for the bell of the tile, which is given sufficient fall so that no water may stand long in it. The soil is packed under the tile and over it, as well, to a depth of 2 or 3 feet as the dam is being built. A foundation of cement and stones is built under the T which is to take the upright inlet tile which usually is set from 8 to 10 feet out or up from the finished dam. All joints except the last on the upright section are cemented and calked to keep the water in the tile. The upper horizontal end of the T may be closed with a vitrified or concrete block or connected to a tile for draining the gully. The outlet end of the tile is carried well out through the lower side of the dam. It is laid in concrete with an ample stilling pool and an apron of concrete or stones. Erosion at the lower or outer end must not be permitted; otherwise, the entire structure is undermined and ruined.

The dam is usually built with teams and slip scrapers. The soil is taken well upstream from the upper side of the location of the finished dam. The soil is packed considerably as the dam is being built. If the soil is too dry to pack properly, sprinkling is desirable, if possible at not too great expense. Allowance of 10 to 12 per cent for settling is advisable. The top of the finished and settled structure should be not less than 4 feet wide or less than 1 foot above the sides of the gully or ravine. The top of the inlet is held down to a distance of 3 feet below the top of the dam (Fig. 135).

Ample spillways should be provided to protect the dam from overtopping during extremely heavy rains. If vegetated spillways are used, these should be sodded upon completion of the dam. Covering the entire outer side and top of the dam with sod is desirable but may require too much labor. Using rye grass and small grain as a seeding mixture on the dam gives protection very quickly. It is advisable to give the outer side of the dam a much flatter slope than is ordinarily done. There is, then, less washing; it is easier to establish a turf on it by seeding; and, owing to the broad base, the dam is safer in every way. The upper side

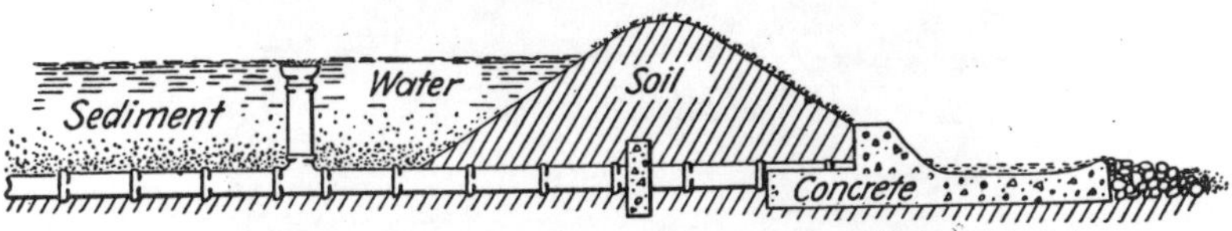

FIG. 135.—Plan for small soil-saving dam.

may be built at the angle of rest for the material used because the catchment basin will soon fill up in any case.

Corrugated, galvanized-iron pipe may be used instead of vitrified tile, but for farm use and from the standpoint of permanence the tile has decided advantages. For larger dams and larger watersheds monolithic concrete inlets are required.

In Wisconsin where rather large ravines are controlled by earth dams two tiles are used, one on each side laid on the solid undisturbed soil.[1] There is thus no danger of the tile's failing because of getting out of line.

A funnel at the top of the inlet tile increases its efficiency to some extent. Ramser suggests the use of woven wire around the top of the inlet to prevent clogging of the tile.[2]

Drop-inlet Highway Culverts.—A wide extension of the use of drop-inlet highway culverts might be made to the distinct advantage of farm land. Drop inlets serve exactly as do soil-saving dams. Full cooperation between landowners and local and state highway authorities is needed. Erosion in gully bottoms above

[1] BATES, C. G., and O. R. ZEASMAN, Soil Erosion—a Local and National Problem, *Wisconsin Res. Bull.* 99, pp. 81–85, 1930.

[2] RAMSER, C. E., Gullies; How to Control and Reclaim Them, *U. S. Dept. Agr., Farmers' Bull.* 1234, pp. 25, 26, 1935.

highways can be limited by the location of the top of the intake of the drop inlet. The cost to the highway of making these inlets effective is very slight. Ample capacity of the highway culvert or sluice is the first requirement for safety of the roadbed, but this is not influenced by the inclusion of the drop-inlet idea in road construction.

FIG. 136.—Concrete soil-saving dam. The underlying soil material here erodes with extreme ease. The following year this dam washed out and fell over into the gully. Photographed by the writer in Mason County, Illinois, 1912.

From thirty to forty years ago in the Midwest roads were washed out by all heavy rains because of inadequate capacity of the culverts. Similar washouts occurred on arterial state highways in south-central New York during the flood in July, 1935, which was unprecedented in that area. Such occurrences are common in many sections that are subject to torrential rains.

Information has been collected on which the required cross section of sluices based on such factors as rainfall intensity, area of watershed, soil, land cover, and slope has now been determined. Ramser,[1] Bates and Zeasman,[2] and Ayres[3] have published such tables.

As new highway sluices are built, the inclusion of the drop-inlet soil-saving dam idea deserves thorough consideration. It appears to be sound soil conservation and good permanent public policy.

Masonry and Concrete Dams.—Dams of masonry and concrete may be used in exactly the same way as described for the earth dam. Failures of these as well as of earth dams have been frequent owing to faulty design and construction. Too often these dams have been built without the advice and assistance of an experienced engineer. All large dams should be planned by and built under the supervision of an engineer who has had some experience with them. Because such dams involve a large cash outlay, they should be so planned and built as to be permanent (Fig. 136).

Federal and state publications have covered the engineering phases of these structures so well that it does not appear wise to include them in detail here (Figs. 137 and 138).

Check Dams.—Any of these types of dams may be used in the accepted sense of check dams. In fact, soil-saving dams are simply large check dams.

Woven-wire Check Dams.—Woven-wire check dams may be used as temporary aids in the establishment of vegetation for the permanent control of erosion. The ends of the woven-wire fencing are extended well into the sides of the gully so that water cannot go around the dam. Straw, leaves, or other slow-rotting material is used above the dam to catch the water and filter out its load of silt and sand. Ample protection in the form of an apron must be provided below the dam; otherwise it may actually increase erosion.

Log Check Dams.—Log check dams have been widely used during recent years, often with excellent results. Their best use probably is in connection with vegetation. If the logs are

[1] *Ibid.*

[2] *Op. cit.*, p. 84.

[3] AYRES, QUINCY C., Recommendations for the Control and Reclamation of Gullies, *Iowa Eng. Exp. Sta., Bull.* 121, p. 63, 1935; also, "Soil Erosion and Its Control," *McGraw-Hill Book Company, Inc.*, New York, 1936.

properly preserved, log dams stand for some years, long enough under favorable conditions for the vegetation to complete the

FIG. 137.—Concrete check dam. In soils underlaid by incoherent sand it is almost impossible to extend the ends of dams into the solid sides far enough to make them safe. The apron has already been undercut. Vegetation may be of real value in gullies of this nature. Photographed, Illinois, 1913.

FIG. 138.—An old type of soil-saving dam. Buttresses appear to be needed although this dam has stood for several years. Photographed, Illinois, 1913. (*Courtesy of R. R. Wells.*)

stabilization of the gully. The quantity of timber required for a series of log check dams is great. Supplying the necessary logs has robbed farm woodlots of timber badly needed for other purposes (Fig. 139; also Figs. 92 and 128).

Concrete and Masonry Check Dams.—Concrete and masonry check dams are effective when rightly designed and substantially built. The advice of an extension agricultural engineer is always desirable in planning these structures. Some of the work can be done with the regular farm help, but competent supervision is good economy.

Fig. 139.—Woven-wire check dam. Note the stones serving as apron below the woven-wire dam as well as below the log check dams farther up the gully. Vegetation is essential in addition to these dams. (*Courtesy of U. S. Soil Conservation Service, High Point, North Carolina.*)

Special attention is directed to the absolute necessity for the extension of both ends of a dam several feet into the solid soil on both sides of the gully to prevent seepage around the ends. An ample spillway must be provided to prevent washing around the ends of the dam. An apron of sufficient size must be provided at the base of the dam to prevent side- and undercutting. Many an otherwise good dam has been lost because of inadequate protection against any undercutting below it. Stilling pools are commonly used in connection with the aprons. Reinforcement or buttresses or both are needed on dams more than two or three feet high. Buttresses are more essential on long than on short ones. The bottom of dams should be placed on impervious material rather than gravel, because of the danger of undermining owing to seepage through the porous material.

CHAPTER XIV

CONTROL OF WIND EROSION

As previously pointed out, man's use and management of forest and soil resources are largely responsible for present soil-erosion conditions in the United States. Destructive wind erosion occurs in three general areas or conditions in this country. The largest area affected is the Great Plains, particularly the drier and sandier parts. All exposed sandy lands on sea and lake shores, river terraces and some other isolated sandy areas, and certain cultivated muck areas are subject to damage by wind erosion. The latter have the highest acre value both of soil and of agricultural products. Owing to differences in soil, rainfall, and evaporation and in cropping conditions, each section requires its own individual treatment.

CONTROL OF WIND EROSION ON THE GREAT PLAINS

As a continuous cover of vegetation is of prime importance in the control of soil erosion by water, so also is a complete vegetative cover of the soil of utmost importance in the prevention of destructive erosion by the wind. Originally nature provided vegetative protection for the soil against both water and wind. It is the part of wisdom for man to study, observe, and imitate nature. As nature clothed the plains with grass, so should man reclothe the areas subject to severe soil blowing—more especially if erosion by water is an added problem. Areas that suffer from erosion year after year had best be returned to permanent grass for pasture or for hay production. The native grasses are most effective in preventing wind movement of the soil. Reestablishment of buffalo grass (*Buchloë dactyloides*), blue grama grass (*Bouteloua gracilis*), and western wheat grass (*Agropyron smithii*) is accomplished with difficulty and very slowly once they have been destroyed by plowing. Nelson[1] suggests seeding crested

[1] Nelson, A. L., Soil Erosion, Archer Field Station, *Wyoming Agr. Exp. Sta., Bull.* 208, p. 31, 1935.

wheat grass (*Agropyron cristatum*) and brome grass (*Bromus inermis*) or slender wheat grass (*Agropyron pauciflorum*) or one or more of these with sweet clover, probably *Melilotus alba.* If the grasses are likely to be covered by drifted soil, he suggests brome grass and western wheat grass as most likely to survive. Alfalfa controls blowing while it is productive, but it leaves the soil dry and pulverulent and subject to blowing. On somewhat less erosive lands, close-growing crops, such as small grains under favorable conditions and Sudan grass and the sorghums, aid greatly in controlling soil movement by winds.

Residues from Crops.—Maximum use should be made of the stubble and stalks and other residues from crops. The burning of straw and stubble as formerly practiced is *positively harmful.* Not only does it destroy the organic matter in the straw or stubble, but it destroys some organic matter in the soil under the fire as well. Soils in areas of low rainfall need every practicable addition of organic matter. Organic matter helps to keep the soil in condition to take up water and at the same time to hold the water against evaporation. And, in addition to all the usual benefits of organic matter in soils, the organic matter in the form of stalks, straw, and stubble helps, when left on or partly on the surface, to reduce the movement of soil by the wind. Standing stalks and stubble check wind movement and thereby reduce the loss of water as snow and by evaporation and the loss of soil by blowing (Figs. 140 and 141).

Finnell reports that the use of crop residues increases the rate of water absorption and lessens evaporation losses about 12 per cent.[1] Such data are, indeed, heartening to those of us who have long urged the marked benefits of adding organic matter to the soil.

Cover Crops.—Cover crops may be seeded for the purpose of protection against the action of the wind. Some consideration of the available supply of moisture may be essential, because a growing crop takes moisture from the soil. If the soil is protected by a mulch, in the event that no cover crop is planted, a net saving of moisture probably results from not planting a cover crop. But a saving of soil attends the growing of the cover crop. The one saving may be balanced against the other depending on

[1] FINNELL, H. H., *6th Southwest Soil and Water Conservation Conference Rept. Proc.*, Oklahoma Agr. and Mech. Coll., p. 53, 1935.

water as the critical factor. The retention of additional water by the cover crop in the event of heavy rains favors the seeding of it.

FIG. 140.—Beach grass holds sand. This beach grass is holding the sand on the barrier dune along the Atlantic Coast of Long Island, New York.

FIG. 141.—Prickly pear holds sand. These prickly pears are found growing abundantly on sand dunes in central Illinois. Photographed, 1911.

The retention of water itself, by keeping moisture in the soil, prevents soil movement by the wind. For this reason and for the benefit of crops, considerable work is warranted for the purpose of holding water in the soil. Level terracing, contour listing, or contour stirring with the duckfoot cultivator all help in holding the water that falls. For this purpose the level terrace with closed ends is used. Owing to the normal shortage of rainfall in the Great Plains, every feasible means of conserving water is beneficial to crops and helps to reduce soil movement.

Pulverized Soil Objectionable.—Under these dry-land conditions a cloddy soil is preferable to a finely pulverized one. A finely pulverized soil of the dust-mulch type is ideal for wind movement. The wind can readily pick up and move the individual particles and the tiny granules. In a cloddy soil, on the other hand, the lumps are too large to be moved by the wind, and the large and small clods protect the fine material against wind movement. Moreover, a cloddy soil takes up rain water to better advantage than does a finely pulverized one. The latter is readily beaten into a fine mud which seals the soil pores and coats over the surface. This makes conditions ideal for loss of water as runoff. A cloddy soil, on the other hand, encourages the absorption of water. Loss of runoff water is serious on a soil that needs every drop of water that falls on it. Implements such as the peg-tooth and disk harrows, which pulverize the soil, are undesirable for use in dry areas. The spring-tooth harrow may be used on hard soil but is likely to do undesirable pulverizing on soils in good tilth. The moldboard plow should be used sparingly, if at all, in dry areas. So-called surface stirring is preferable to fairly deep plowing. The duckfoot cultivator and the lister produce the most desirable soil condition for the Great Plains area, the soil of which is subject to wind action (Fig. 142).

Keeping Soil Rough Helpful.—Making lister or duckfoot cultivator furrows at right angles to the prevailing wind direction is good practice.[1] Soil particles blown from the interfurrow spaces drop into the furrows and are held there. Further movement is checked until the furrows are filled. New furrows are then made in order to continue the protection for the soil. Owing to the shifting of the wind, occasional cross furrows should be of some additional service (Figs. 143 and 144).

Crops are planted in furrows instead of on the level surface, as is the practice in humid areas. Corn is planted in lister

[1] THROCKMORTON, R. I., and F. L. DULEY, Soil Fertility. *Kansas Agr. Exp. Sta., Bull.* 260, p. 30, 1932.

furrows, and small grains in furrows made by a special furrowing grain drill. Crops are planted with the rows at right angles to

FIG. 142.—The duckfoot cultivator. The duckfoot cultivator produces a cloddy rather than a pulverulent condition of the soil. This condition favors control of blowing as do also the furrows made by this cultivator. (*Courtesy of International Harvester Company, Chicago, Ill.*)

FIG. 143.—The lister. Note the depth of furrows produced and the cloddy rather than pulverized condition of the soil. This favors control of erosion. (*Courtesy of Deere and Company, Moline, Ill.*)

the prevailing wind direction. Not only is the soil thus protected before the grain comes up, but it is protected by the rows of grain during its growth and by the rows of stubble after the grain is harvested, all of which are at right angles to the prevailing winds. These furrows aid in the absorption of rainfall and control soil movement by the wind.

Leonard and Stewart recommend strip listing for fields in which the wheat failed to germinate in the fall. The strips may be a single furrow, or more than one, spaced 50 to 100 feet apart. The blowing soil is caught in the furrows, especially that carried near the surface of the soil by saltation.[1]

FIG. 144.—Lister furrows. Lister furrows hold snow and when made on the contour they hold water as well. The duckfoot furrows act in this way, too, but to a lesser extent owing to their being shallower.

Both types of furrows aid greatly in conserving moisture and in holding the soil against blowing. *(Courtesy of Deere and Company.)*

Strip-crop Rotations.—Crops vary in their ability to hold soil against wind erosion, just as they vary in the extent to which they control or facilitate erosion by water. Corn, potatoes, cotton, fallow for wheat, and beans permit easy wind movement of the soil; the sorghums, sudan grass, small grains, and native grasses and alfalfa and sweet clover, to a lesser degree, resist wind erosion. Crops resistant to wind erosion are grown in alternate strips with such crops as corn, potatoes, or beans. As with furrows, so the strips are laid out at right angles to the prevailing wind direction in order to obtain maximum protection. Nelson recommends a two-year rotation of corn and "spring-drilled"

[1] LEONARD, C. D., and T. G. STEWART, Keeping the Farm at Home, *Colorado Agr. Ext. Circ.* 97-A, p. 6, 1935.

crop.[1] In that area crop rows are run north and south for best protection against westerly winds. He states that on fairly good wheat land containing considerable clay soil, blowing is largely controlled if cropped in strips from 15 to 20 rods wide. On sandy land the strips are narrowed to from 8 to 15 rods in width. A spring-drilled crop is alternated with corn on either strip width. He suggests a five-year rotation for strip cropping with the following crop sequence: first year, corn; second, spring-drilled crop; third, rye; fourth, fallow; and fifth year, winter wheat. The spring-drilled crop should mature as early as possible in order to give the soil time to accumulate moisture for the rye. The stubble from the spring-drilled crop and the rye is left on the surface when the land is fallowed with the duckfoot cultivator, and, except on sandy land, soil blowing should be prevented by it and by the wheat during the winter. This rotation can be repeated in 16-rod strips, two of each crop, on an area 160 rods in width, or on 160 acres.

Brandon, quoted by Leonard and Stewart, recommends a strip-crop five-year rotation based on results obtained on the U. S. Dry-land Station at Akron, Colo. The general principles governing the plans on the Archer Wyoming field apply here, also. Forage millet, proso, and cane are added to the Wyoming crop list. The substitution of sudan grass for the millets is a possibility. The corn and beans may be seeded in alternate strips. Some of the forage crops including sudan grass may be grown in strips across and around the fallow land for protection against destructive blowing during the summer. Listing the bean land immediately after harvesting reduces blowing. The regular rotation provides stubble crops immediately preceding fallowing to help in the control of blowing during the fallow year.

The sorghums are of first importance in the central and southern parts of the Great Plains, from the standpoint both of feed production and of the control of wind erosion. McGinnis rates the stooling grain sorghums as very effective in holding the soil against wind action.[2] The fibrous stalks and the large thick root

[1] NELSON, A. L., Soil Erosion, Archer Field Station, *Wyoming Agr. Exp. Sta., Bull.* 208, pp. 24–29, 1935.

[2] MCGINNIS, B. W., Utilization of Crop Residue to Reduce Wind Erosion, *U. S. Dept. Agr.; The Land, Today and Tomorrow*, Vol. 2, No. 4, pp. 12 to 14, 1935.

systems of milo maize, hegira, kafir corn, cane, and sudan grass are so durable as to hold the soil from season to season. He reports that milo maize held the soil in the worst windswept section from the autumn of 1933 to the spring of 1935.

Windbreaks.—Wherever trees and shrubs can be established, species adapted to the locality can be used to help reduce wind erosion. The protected distance varies with the height of the windbreak. Its effectiveness depends on its density. At the Archer station in Wyoming it is estimated that the distance protected to the leeward of the trees was from 20 to 35 rods, or from 330 to 577 feet. On this basis one windbreak running through the middle of a square 40-acre field transverse to the wind direction should aid materially in reducing soil blowing and the loss of water by evaporation from plants and soil as well.

Albert reports that a windbreak 6 feet high in Wisconsin planted in 1928 gave protection to the leeward of about 8 rods, or 132 feet.[1] Owing to the shifting of the wind, protection from the southwest winds in Wisconsin is not sufficient. Secondary windbreaks at right angles to the primary ones may be needed even though they do interfere with tillage operations at times. He reports on the use of lattice fence but owing to the shifting of the wind does not regard such obstructions as being of great value in Wisconsin.

Annual plants such as the mammoth Russian sunflower and sudan grass are of value as windbreaks. These crops are particularly effective in catching and holding snow which if taken up by the soil as water is of real value to the next season's crop.

CONTROL OF BLOWING ON PEAT LANDS

Peat lands, usually called muck in this country, owing to their method of formation, are almost universally located on low areas, and often they are surrounded by trees. These help greatly in checking wind erosion on small peat areas. On some of the larger areas, however, damage to tender young vegetable plants is severe in some seasons. The acreage injured is not always great, but the value of the crop is normally high. And if, for example, a crop of lettuce is killed, not only is the labor

[1] ALBERT, A. R., Coping with Sandstorms, *Wisconsin Agr. Ext., Stencil Circ.* 14 B, 1934.

of getting it started lost, but the next crop is delayed in reaching market and may suffer in price as a consequence.

In the larger peat areas in New York, willow windbreaks were planted some years ago, some of the older ones now being from 40 to 50 feet high. Such large windbreaks are so wide as to occupy many acres of valuable crop land. Yet the owners appreciate the good work that the willows are doing. On some newly developed areas the willows are only 10 to 15 feet high. Some damage from blowing takes place because the willows do

FIG. 145.—Willow windbreak on peat. This old willow windbreak aids materially in the control of erosion on this peat area in western New York. Photographed, 1931.

not yet afford complete protection. As temporary aid, picket fence is used during the early critical part of the windy season. If blowing is unusually severe, old burlap bags are hung on the lower part of the fence to check air movement near the soil (Figs. 145, 146, and 147).

Strips of rye of drill width or less are seeded in the fall and allowed to grow. Considering its height (4 to 6 feet on heavily fertilized peat) rye is effective in checking wind movement on peat lands.

Rolling of peat has been resorted to for packing the dry surface material so that it takes up water by capillary action. Because only dry peat is moved by the wind, rolling may be effective, provided the peat immediately under the surface contains a good supply of water. The early cultivation of small vegetable

FIG. 146.—A cover crop of wheat following lettuce. This cover crop protects the soil, and conserves plant nutrients. Note the willow windbreak at the opposite side of the field. Photographed, 1932.

FIG. 147.—Fences used as windbreaks. These picket fences in addition to the willow windbreaks afford a fair degree of protection to crops against the blowing of peat. In extreme winds some damage is done in spite of the windbreaks, especially if these are spaced too far apart.

plants is so shallow as not to be of much service in the way of roughening the surface of the peat for the control of blowing.

CONTROL OF THE BLOWING OF SAND IN HUMID AREAS

Unprotected sand is subject to movement by the wind regardless of where it is located. In nature, most of it was held in place by forest growth, but some of it, as in central Illinois and on the Great Plains, was held by grasses. On the seashore and along the larger lakes, sand has been migrating for many years, in fact may never have been stabilized owing to wave action, severe winds,

FIG. 148.—Blowing sand is reforested with difficulty. This young pine tree may win its contest with the wind, but its chances are not very good. Ten inches of sand has blown out from under it since it was planted. The trees have been blown out of the bare sand. Photographed, 1932, in northern New York.

and the detrimental effects of salt water on ordinary types of vegetation.

Reclothing blowing sand with forest growth is the ideal means of anchoring migrating sands. In places the movement of the sand is so pronounced that the seedlings are blown out (Fig. 148); that is, the sand blows out from under them. In such situations, reforestation is accompanied with some difficulties (Fig. 149). Windbreaks can be used for a period until the trees get a sufficient foothold to check the movement of the sand. Another method is to plant the tree seedlings; cover the surface of the sand with straw, leaves, pine needles, or fine brush; and lay brush over the mulch to hold it in place. The brush-held mulch prevents much sand movement until the trees have become sufficiently well established to anchor the sand. The black locust (*Robinia Pseudo-acacia*) serves well in such places (Fig. 150). After being thoroughly inoculated it makes good growth on dry sandy soils. In their habitat conifers may well be mixed with or, better perhaps, planted in alternate strips with locusts. The conifers,

FIG. 149.—Pine stump tells a story. This pine stump which stands a few feet from the pine shown in Fig. 148 grew to maturity before man disturbed the forests. About three feet of sand has blown out from under what was once a vigorous growing tree.

however, require protection over a longer period than do the locusts because of their relatively slow growth.

Wherever, as in the Midwest, the sand is covered with a sparse growth of grasses and prickly pears (*Opuntia polyacantha*), locusts may be established without any special protection. The pitch pine (*Pinus rigida*) and some others have been planted, but insufficient time has elapsed to show whether the pines can succeed in dry sand under the high temperatures of the Midwest. Various oaks (*Quercus* spp.) and the shagbark hickory (*Carya ovata*) are growing naturally in that sand area. These species, therefore, can be grown on that sand in addition to the locust.

The long-leaved reed grass, big sand grass, or carizzo (*Calamovilfa longifolia*) is growing in actively blowing sand in central Illinois and is found in adjacent states from Michigan to Alberta and westward to Colorado. It spreads by means of strong, scaly, creeping rhizomes and makes good growth in the fresh sand at the immediate leeward edge of "blowouts." Planting trees in the

FIG. 150.—Stabilizing a "blowout." A farm boundary crosses this blowout. The sand in the immediate foreground was planted to black locusts, seeded to rye, and the surface covered with a light dressing of straw. Hedge branches are scattered about over the straw to hold it in place. No sand movement took place on the treated area, but sand was being blown from the untreated sand of the noncooperator in April, 1936. This blowing in of sand may defeat the purpose of the treatment.

grass around a blowout and then planting the blowout itself to this grass and locusts with ample straw or litter for the necessary temporary protection brings the sand under control in time. The locusts may be planted very close together, as many as five to ten times the usual number of trees being used to the acre. Not all of the seedlings will survive the unfavorable conditions, but the close spacing enables the locusts to cover the sand quickly and finally to anchor it (Fig. 151).

During the past twenty-two years the writer has observed black locusts planted as sprouts on blow sand in central Illinois between twenty-five and thirty years ago. The fires that have gone over the area about once in three years on the average produce such intense heat as to kill much of the aboveground growth. Sprouts,

FIG. 151.—Long-leaved reed grass. Long-leaved reed grass (*Calamovilfa longifolia*) is growing in the most actively blowing sand. Transplanting this hardy grass in the most actively blowing areas aids the black locust in completing the stabilization of sand in central Illinois.

however, come up from the live roots and thus maintain protection for the sand. This planting has spread by means of suckers until the locust grove or thicket covers a much larger area than that originally planted.

By the spring of 1934, the locusts had covered the sand with leaf mold to the extent of more than a ton of dry matter to the acre since the most recent previous fire in 1930. Even though an extremely severe fire went through this locust plantation in the spring of 1936, the sand nevertheless has been held in place by the windbreak effect of the old top growth and the sprouts from the locust roots.

In this section it is notable that Kentucky bluegrass (*Poa pratensis*) comes in under locust trees on blow sand, but the bluegrass is not found elsewhere on the sand. The bluegrass comes in as a result of the lowering of the temperature by the partial shade, the addition of nitrogen in the locust leaves, and the additional water held by the locust-leaf mulch. Such grass is a material aid in keeping the sand stationary.

Wild Legumes.—Annual wild legumes including the trailing wild bean (*Strophostyles helvola*) and the partridge pea (*Cassia chamaecrista*) grow well on sandy loams and loamy sands. The partridge pea grows on pure sands, also. Both plants add nitrogen and organic matter to the soil and hold sand physically, as do nonlegumes. These legumes can be spread over sandy land by seeding them. Some years ago the writer collected a few ounces of seed of the trailing wild bean, which he scattered over some 60 acres of loamy sand and sandy loam. The entire area is now well covered with this plant; in fact, the trailing wild bean is a physical handicap to corn plants. It is of great value, however, from the standpoints both of supplying nitrogen and organic matter and supplying physical protection to the soil against blowing.

Windbreaks.—Windbreaks of black locust, osage orange (*Maclura pomifera*), or other shrubs and small trees are useful in reducing the movement of sand on the lighter loamy soils of the Midwest. Planting windbreaks on each 80-rod line and midway between the north and south lines helps materially in reducing soil movement by the wind. It is recognized, of course, that such windbreaks, particularly those through the middle of 40-acre fields, are objectionable because they interfere with large-scale tractor tillage operations.

Seeding at Right Angles to Prevailing Wind Direction.—Seeding small grain and cowpeas (*Vigna sinensis*) and soybeans (*Soja max*) at right angles to the prevailing southwest winds aids materially in checking sand movement. Owing to the shifting of the wind from day to day, seeding an occasional drill width of a crop at right angles to the general direction of seeding, checks sand movement along the rows when the wind shifts from its prevailing direction. Cowpeas are particularly sensitive to the abrasive action of sharp sand particles on the tender stems of the young plants. Such damage kills the crop and necessitates replanting once, and in some seasons, a second time, in order to obtain a stand. May and June are the critical months for wind damage to this crop.

Small grains, especially the winter grains wheat and rye, should be seeded transversely to the prevailing wind direction. Seeding a drill width or two crosswise every 50 to 100 feet is probably worth while, particularly in the winter grains which are exposed

FIG. 152.—Plowing double furrows checks sand movement. These double furrows, which were spaced from 12 to 15 feet apart, control sand movement. Any sand moved on the inter-furrow space drops into the furrows until these are filled. When this occurs the farmer makes new furrows and thus controls the loss of soil. Photographed, April, 1936, near Bath, Ill.

to wind action throughout winter and early spring. Insufficient snow cover permits considerable winter damage on exposed knolls of light soils unless precautions are taken for controlling wind erosion.

Furrowing to Check Blowing of Sand.—Furrowing with an ordinary turning plow is practiced for preventing loss of sandy soil. Two furrows are thrown together at intervals of about 15 feet depending on the danger from blowing. Cross furrows may be made if the wind shifts materially, and in any event new furrows are made when the old ones have been filled with sand. This method of controlling sand movement is probably too expensive for general use. Some means of vegetative control or a covering of manure is preferable except, perhaps, in vegetable production (Fig. 152).

CONTROLLING THE BLOWING OF BEACH SANDS

The blowing of beach sand is an old problem, one that is in need of still more attention. Vegetation must be relied on mainly,

FIG. 153.—American beach grass on Long Island. The American beach grass (*Ammophila breviligulata*) has been planted on hundreds of acres of blowing beach sand on the Atlantic Coast of Long Island, New York. The picket fence at the upper left intercepts sand from a large unplanted area of sand. The grass is making a fairly good growth but it needs nitrogen. The trailing wild bean (*Strophostyles helvola*) in the foreground might be seeded and might aid greatly in the final stabilization of blowing sand.

but windbreak fences check sand movement and aid vegetation in getting a firm foothold. Beach, psamma, maram grass, or sea sand reed (*Ammophila arenaria*) is used for anchoring beach sand on Cape Cod and on the Pacific Coast. American beach grass (*A. breviligulata*) grows on coastal sand dunes from North Carolina to Newfoundland and on the sandy shores of the Great Lakes. These grasses spread by means of hard, scaly, creeping rhizomes. This method of reproduction makes strong growth possible under the extremely adverse conditions of their habitat. Extensive plantations of the American beach grass have been made on the south shore of Long Island for holding the coastal sand dunes (Figs. 153 and 154).

Legumes, shrubs, or trees should be planted among the beach grass to hold the sand after the grass has begun to function. Hardy plants adapted to windy shore lines and subject to the salt

FIG. 154.—Bayberry holds sand in place. Bayberry bushes are effective in holding Long Island beach sands against blowing.

spray of the sea are the ones to use. Suitable evergreens are of special value during the late fall, winter, and early spring. As stressed so often, legumes are needed in this sterile sand. The black locust (*Robinia Pseudo-acacia*) and the shipmast locust may serve this purpose in some areas. In many areas may be found annual and perennial legumes which can be used to real advantage. Not only may they hold the sand directly, but they help the nonlegumes to make a stronger growth and thus afford increased protection. A complete cover of live vegetation is the best means of holding beach sand in place and, therefore, of protecting other land and other property against encroachment and burial by the migrating sand.

Special References

Soil Drifting Control in the Prairie Provinces, by E. S. Hopkins, S. Barnes, A. E. Palmer, and W. S. Chepil, *Dept. of Agr., Dominion of Canada, new series, Bull.* 179, 1935.

Soil Blowing and Its Control in Colorado, by J. F. Brandon and Alvin Kezer, *Colorado Agr. Exp. Sta., Bull.* 419, 1936.

CHAPTER XV

CONTROL OF WAVE EROSION

The control of wave erosion is, in the main, a public problem, one of highly specialized engineering. The cost of shore protection by engineering means is so great as to be practicable only for harbors, cities, summer resorts, and other public properties. The need for protecting farm land is, indeed, pressing at many points on ocean and lake shores. According to Patton, the Atlantic Ocean cut away 2,200 acres of the New Jersey shore during the eighty years from 1840 to 1920.[1] Some other losses of land to the ocean in various places are shown on pages 45 to 46.

The loss of farm land to the ocean and to the larger lakes and rivers, as a result of wave action, is large in the aggregate, yet no one farm fronting on the water loses any considerable acreage in a single season. Erosion on the shores of the Great Lakes is not easily controlled in the more exposed places. Records of definite action for the control of wave erosion on farm lands on lake shores have not come to the writer's attention. Work of that nature, however, has undoubtedly been done.

Here, as in so many phases of soil stabilization, men may advisedly observe and practice nature's methods. In numerous places all of the material that the wind was capable of moving has been carried away, leaving the surface protected by a so-called *desert pavement* (Fig. 155). Water has done the same on millions of acres, and the result is a pavement of stones. From shore lines composed of glacial debris the waves have carried away all of the fine material. Large boulders are left behind, and these actually protect the land from further rapid cutting away (Figs. 156 and 157). The banks of many streams are lined with water-loving trees, shrubs, grasses, and other plants. And relatively stable stream banks are the result.

[1] PATTON, R. S., Problems Involved in Coast Erosion, *Military Eng.*, Vol. 16, p. 463, 1924.

FIG. 155.—A desert pavement caused by wind erosion. This area was once heavily timbered. The land was cleared and cropped. The wind removed the sand leaving the gravel and stones.

FIG. 156.—Erosion on the Atlantic Coast of Long Island. This rock pavement or barrier is insufficient to protect the coastline which is composed of glacial materials. Erosion of glacial soil materials is unusually rapid.

Each situation constitutes a separate problem. The writer's one hope is that the suggestions that follow may prove helpful in protecting farm land and perhaps other shore property from wave erosion.

FIG. 157.—Erosion on the New York shore line of Lake Ontario. This glacial material contains few large boulders, consequently further erosion does little in developing a shore protection of large rocks.

ON LAKE SHORES

Driving piling and placing rocks, mostly 2 feet or more in diameter, on the shore line are helpful expedients, but such undertakings are so costly that few individual landowners can attempt them. Practical control measures within the attainment of the individual owner are essential. It appears that some of the means of control of the larger gullies are applicable in the control of erosion on steep shore lines.

Planting Trees and Shrubs.—Planting willows (*Salix* spp.), adapted to the locality, back of the piling and rocks, or back of the rocks alone, or near the water's edge should provide some protection in a short time. One way of planting willows in such places is to drive willow stakes several feet into the soil near the water. Such stakes are not easily washed out. They take root and soon afford the shore a measure of protection (Figs. 158 and 159).

Wherever the waves strike the shore at an oblique angle, the scouring action may be so severe as to prevent the willows from establishing themselves. In such situations structures are needed, at least for temporary protection.

FIG. 158.—Willows and other trees protect this lake shore. Note the growth of willows near the edge of the water.

FIG. 159.—Native shrubs covering the raw bank. These native shrubs are hardy and capable of good growth on this soil.

Properly handled willow and cottonwood (*Populus* spp.) slips of adapted species may be planted on the bank up from those placed near the water's edge. Still farther up on the bank the black locust (*Robinia Pseudo-acacia*) and, perhaps, redbud (*Cercis canadensis*) may be planted. Legumes are especially desirable owing to their ability, when inoculated, to obtain their own supply of nitrogen and to make rapid growth on raw subsoils, which are usually unfavorable for other trees. The locust, in particular, makes relatively rapid growth in such places (Fig. 160).

FIG. 160.—Coast erosion on the New York shore of Lake Ontario. When wet, this glacial material is peculiarly responsive to gravity.

Because evergreens protect the soil against waves, wind, and rain they deserve wide use. They are highly desirable on steep, dry banks. Red cedar (*Juniperus virginiana*) and hemlock (*Tsuga canadensis*) grow well on many such unfavorable situations.

Various shrubs mentioned on pages 226 to 229 which grow satisfactorily afford protection against erosion and may be used in the localities and on the soils to which they are adapted.

Planting Vines.—The planting of the most vigorous-growing, adapted, native, woody vines is to be encouraged, also. While climbing vines are very useful, those which if planted in the good soil at the top of the bank grow when hanging over it are especially helpful in protecting steep, raw banks. Among the desirable vines may be mentioned Hall's Japanese honeysuckle (*Lonicera japonica Halliana*), Virginia creeper (*Ampelopsis quinquefolia*), American bittersweet (*Celastrus scandens*), and virgin's bower (*Clematis virginiana*). All of these are mainly in the class of climbers. The wild grapes (*Vitis* spp.) make good growth and afford desirable protection when growing on and hanging from the top of the bank.

Planting Grasses and Legumes.—Reed canary grass (*Phalaris arundinacea*), which grows in wet soils, may be planted near the water's edge and on wet banks. Planting this and other moisture-tolerant grasses such as rough-stalked meadow grass (*Poa trivialis*) and creeping bent grass (*Agrostis palustris* Huds.) among or on both sides of the willows should help to protect the soil against wave action and rain. At some points on precipitous shore lines, such as those on the Great Lakes, sloping of the bank may be one of the more important preliminary steps in their stabilization.

Such legumes as white sweet clover (*Melilotus alba*), yellow trefoil (*Medicago lupulina*), partridge pea (*Cassia fasiculata* and *C. chamaecrista*), trailing wild bean (*Strophostyles helvola*), and many others including some locally growing perennials are especially desirable for producing a vegetative cover for these raw banks. The seed should be inoculated before planting in order to obtain the best results. The trailing wild bean owing to its climbing habit possesses some advantages over other legumes.

Grasses, adapted to the locality, to the soil and its moisture content, and to the exposure, should be seeded along with the foregoing legumes. The better sward-forming grasses for each situation deserve preference. The planting of the stolons of some of the grasses hastens sward formation to such an extent that protection may be attained between the expected storm periods. This is hardly possible if seed alone is planted. The use of fertilizers, manure, and perhaps straw as a mulch may be expected to hasten the development of the grass.

ON RIVER BANKS

Similar treatment of river banks may be expected to yield good results. Banks that are being rapidly undercut will require structural protection before they can be stabilized by means of vegetation. Willows and cottonwoods are helpful in controlling

FIG. 161.—Willows border a stream in eastern Kentucky. The willows growing near the water's edge afford much protection to the stream banks.

FIG. 162.—A stabilized ocean coastline. The boulders have protected the land sufficiently for vegetation to gain a foothold. The boulders and a thrifty growth of native shrubs have completed the stabilization of this section of the Atlantic coastline of Long Island.

FIG. 163.—A gravel coastline on Long Island Sound. At the left it may be noted that shrubs are making a beginning in clothing this gravel with a protective cover.

FIG. 164.—Stable coastline on the north shore of Long Island. Practically no erosion is taking place here. Among the plants which are protecting the soil are: bayberry (*Myrica carolinensis*), sumac (*Rhus copallina*), yarrow (*Achillea Ptarmica*), wild rose (*Rosa virginiana*), and red fescue (*Festuca rubra*).

erosion on the banks of the smaller streams. The elms and the sycamore and many other trees grow well on moderately dry stream banks and aid greatly in their stabilization (Fig. 161).

ON OCEAN SHORES

As already indicated, ocean shore protection is regarded as an engineering problem. Some use may, nevertheless, be made of the suggestions for protecting farm land that fronts on the ocean shore. The use of beach grass (page 260) is helpful in stabilizing sandy shore lines (Figs. 162, 163, and 164).

Instead of sand dunes or precipitous banks the shore line in many places, as on Long Island, consists of shallow marshes and mud flats. Smooth cord grass (*Spartina alterniflora*) and its close European relative *S. townsendii* are particularly effective in catching and holding the soil and in adding organic matter. In time, the marshes are built up into dry land. According to Hitchcock, *S. townsendii* is planted along the sea coasts of northern Europe with excellent results.[1]

[1] HITCHCOCK, A. S., Manual of the Grasses of the United States *U. S. Dept. Agr., Misc. Pub.* 200, p. 489, 1935.

CHAPTER XVI

CONTROL OF EROSION ON PUBLIC HIGHWAYS

It is most gratifying to note that soil erosion is recognized by our most progressive highway engineers not only as a grave problem but as one that can be solved in a large measure by the making of more gradual slopes than are common and the planting on them of well-selected vegetation. The results of some efforts to control erosion on highways have been disappointing, but others have been a pronounced success. All of those who recognize this problem and who are working for its economical solution deserve public commendation.

Soil erosion on public roads is an old problem which has been with us since roads were first built in this country. The problem has become more acute since the introduction of motor vehicles and of the present-day power road-building machinery. Higher embankments are being built and deeper cuts are being made than in the day of the slip scraper drawn by horses or mules. With these changes have come a marked intensification of the roadside soil-erosion problem. A cut 40 feet deep, for example, presents far more difficult problems than does one only 5 to 10 feet deep (Fig. 165).

In the past the taxpayer has demanded that the largest possible mileage be built with the funds appropriated. Moreover, it was not until after some years of cumulative erosion on highways that the public became aware of the seriousness of this problem. On the highways of the country are thousands of miles of cuts and fills which are well stabilized by grasses and shrubs, and many other cuts of bare soil on both old and new highways. Some of these cuts and embankments are steep enough for rain to cause considerable washing (Fig. 166). Some of the finer material finds its way into reservoirs, and some of the coarse comes to rest at the base of the slope on the highway and must be removed.

The disposition of highway drainage water is attended with difficulties. In some places it has been turned out on to pastures, meadows, or cultivated fields, and gullies have sometimes resulted (Fig. 167). On the other hand, water from privately

FIG. 165.—A highway cut in eastern Tennessee. Some of this cut is through rock but soil has rolled down over it.

FIG. 166.—Road cut in boulder clay in central Illinois.

owned land has been turned into highways and has led to severe damage. Close cooperation between highway officials and landowners is essential (Figs. 168, 169, 170, and 171).

FIG. 167.—Road water did this. This is a natural drainage course. It requires stabilization. If treated at once a grass waterway can be developed. If neglected a gully which cannot be crossed with ordinary farm implements will develop in a few more years. Photographed, North Carolina, 1936.

FIG. 168.—A terrace outlet caused this erosion. Two small terraces about seventy-five feet wide on a 5-per-cent slope drain into the highway ditch. Early action is needed in cooperation between the farmer and the road authorities. Photographed, North Carolina, 1936.

FIG. 169.—Gully caused by water along road in South Carolina. The drainage from about 25 acres was conducted into the road ditch and carried down parallel to a railway at right. Slope of land is about 5 per cent. Photograph taken near Spartanburg, S. C.

FIG. 170.—Head of gully shown in Fig. 169. The water was carried across the roadway in the two tiles shown. The owner is seated on the emergency guard rail. Early action is needed to save the highway.

Many a mile of new or recently built highway, especially in the South, is placed alongside of the old one, largely because of the lowering of the old roadbed through the removal of soil by erosion (Fig. 172). Because the old roadway was in a depression, no drainage could be accomplished except down the sides of the road. It was found to be advisable to relocate the road up on the level of the adjacent farm land where drainage might be provided with much less difficulty. In other places erosion has caused

FIG. 171.—Discharge of water by a steel road pipe. If this water does not fall on rocks, brush, concrete, or other stable material undercutting of the road is inevitable. Photographed in central Illinois in 1914 following a shower.

severe damage to the highway and has entailed heavy public expense (Fig. 90).

Erosion Damage on Roads in New York.—In seven central-southern New York counties a study was made of highway maintenance costs. As reported by De France, the cost of repairing erosion damage on state highways in these seven counties was $30,000 a year.[1] By including the county highways this amount was raised to $60,000. Moreover, this amount on the state highways was one-half of the total state-highway maintenance

[1] DE FRANCE, J. A., "The Control of Erosion on Embankments by Vegetation," *U. S. Soil Conservation Service, Ithaca, N. Y.*, 1935.

FIG. 172.—A road in a depression in deep loess. Erosion has lowered this roadway. No outlet for water is possible, except down the highway. Photographed in western Illinois, 1912.

FIG. 173.—Lake-laid soil in New York, unstable when wet.

costs in these seven counties. On the basis of the studies made, it is estimated that the normal annual cost of repairing erosion damage to New York highways is $500,000 a year (Fig. 173).

Wide Cooperation Essential.—The Federal Bureau of Public Roads is cooperating with the U. S. Soil Conservation Service in a study of highway erosion problems and the relationship of these problems to soil conservation on adjacent farm lands. The wholehearted cooperation of the highway-construction engineers and of all other public-road authorities, landscape gardeners, agronomists, and men well versed in the physics and mechanics of soils is needed. The problem is one of such magnitude as to engage the best efforts of all of these groups in its solution.

Pending the results of the studies already undertaken, it may not be out of place to set down provisionally some of the things that in the light of present information can be done to advantage.

SOME SUGGESTIONS FOR THE SOLUTION OF HIGHWAY EROSION PROBLEMS

The following suggestions are based partly on the observations on many miles of highway cuts and fills in different states and what has happened to them; on the behavior of different soils under various moisture conditions; and on the effects of grasses, shrubs, and trees on various slopes.

The Slope of Cuts and Fills.—Here, as in many phases of soil-erosion control, nature points the way. Slopes in nature tend to be gradual and to be covered with a complete protection of vegetation. That is a good starting point. Soils and the underlying material vary greatly with respect to stability on different degrees or percentages of slope. A dry gravel, for example, stands moderately well at the angle of rest. Some sand and pebbles are loosened by wind and rain and roll down the slope. Such porous material absorbs most of the rain falling on it, but any water that is not absorbed flows rapidly down such steep slopes and carries much sand and gravel into the roadway. Handwork is often involved in its removal.

At the other extreme are lake-laid or other water-bearing soils. The water need be in the soil only at certain seasons of the year to cause much trouble. In order to eliminate slides which now occur frequently, the slope of cuts in such soils may properly be changed from the usual 1 to 1, or a 100-per-cent slope, to 2 to 1, or a 50-per-cent slope. In case of extremely unstable soils a somewhat flatter slope may be necessary. Such slopes appear extremely flat, but nature by means of frequent slips and slides is tending in the direction of gradual slopes (Fig. 174). A construction problem is involved, whether it is not desirable to give cuts their probable stable slope at the time when the road is being built. The work can then be done with the usual heavy power machinery, whereas slides, in addition to causing traffic delays,

Fig. 174.—Slip erosion on Dunkirk, a lake-laid soil. The slope of the field at the slip is 38 to 40 per cent. Photographed, March, 1936.

involve the use of small machinery and hand labor, sometimes at considerable expense.

The function of the experienced soils man is to assist the engineer in deciding upon the percentage of slope at which any particular soil material will stand without serious slipping. It is realized that deep cuts require wide right of ways which cost the public considerable amounts of money. Many steep cuts have slipped and carried down the farmer's fence and eventually a considerable strip of land for which the public has not paid the landowner. It is, of course, not the intention of the public to take the property of any citizen without properly compensating him for it (Fig. 175). The point is that if the cut is made steeper than its angle of rest, nature finally reduces it to that angle.

Relatively Flat Slopes Desirable.—High embankments made for overhead crossings or fills in deep ravines, in some cases, might be made rather wide at the base and with correspondingly gentle slopes. In order to do this, right of ways must be wider at some increased public expense. On the other hand, a 1½-to-1 grade, for example, is a slope of 66⅔ per cent, one on which erosion is almost certain to occur. Even after stabilization it is difficult to hold so steep a slope under much concentration of water from any cause. Compared with cuts, however, embankments have led to relatively little difficulty.

FIG. 175.—A slip on a road cut in southern Ohio.

Diversion Ditches.—A road cut at the base of a long slope requires protection during the process of growing the permanent vegetative cover. This may be accomplished by making a shallow diversion ditch a few feet above the upper edge of the cut. Such a ditch should be wide and have a flat bottom which must be protected from erosion by the water that it intercepts. In the past, some diversion ditches have been of the narrow, V-shaped type without protecting vegetative cover, and erosion has followed (Fig. 176). On the steep slopes, tall coarse grasses are needed in the ditches to check the flow of water sufficiently to reduce erosion. Checking the velocity reduces the carrying capacity of the ditch. For this reason, ditches that are to be seeded must have cross sections of greater area than is required otherwise. Unless legumes are included in the seeding mixture, manuring or fertilization of the grass is needed to produce the desired protection.

Disposition of Road Water.—Water inevitably collects in the ditches along roads on slopes. Letting the water out at short intervals is desirable in order to avoid the concentration of much water in one place. This is because of the well-known increase in

FIG. 176.—Road-cut diversion ditch. The road-cut diversion ditch is a good device. In this case water from a field road was turned into it. The additional water caused this erosion. Slope about 12 per cent. Photographed in North Carolina.

eroding and carrying power of water in large quantity and of high velocity. Proper spreading of the water on timbered areas or on a thick, erosion-resistant sward of pasture or meadow grass which is well stabilized is essential.

On secondary roads, particularly, the drainage ditch at the sides usually is the ordinary country subsoil material which often is subject to severe erosion. To prevent erosion in highway ditches, check dams of concrete, stone, or creosoted wood have been used with fair success; but a concrete, asphalt, brick, or stone gutter appears to give the best protection (Fig. 177). Tile with drop inlets may be used to advantage under some conditions.

Silt basins or sediment wells are needed on such drains at points where the slope is materially reduced.

Vegetative Stabilization of Road Slopes.—Vegetation cannot be expected to hold soil or slopes that are steeper than the angle of rest under the most unfavorable moisture conditions. Sloping banks to the proper angle is the first step. Rounding off the slope at the top helps to some extent.

FIG. 177.—Creosoted plank check dams in central Illinois. Slope of road ditch 6 per cent. Data supplied by T. R. Wire of the U. S. Demonstration project at Le Roy, Ill.

Cuts and fills may be kept from washing by establishing a cover of grasses and legumes, vines, shrubs, and small trees. Large trees and the taller shrubs are held down by cutting or pruning, as may be necessary. In places, all of these forms of vegetation used together may give the quickest and best protection to steep highway slopes.

Grasses and Legumes.—If a single seeding mixture is to be used in an area such as an entire state, a number of grasses and legumes must be used to produce a suitable protective covering under a wide range of conditions with respect to soil, exposure, and moisture. The plants that have value on dry gravelly banks can make little growth on heavy, seepy soils. Legumes are needed in mixtures, because the slopes usually consist mainly of subsoil material which is very low in, if not almost devoid of, organic matter and nitrogen. The application of a light coat of productive surface soil to a slope usually helps in producing a suitable sward for protection against erosion. Using 800 pounds to the acre of a good mixed fertilizer, such as a 5-10-5[1] or 4-16-4, with or without coarse manure, hastens the growth of the grass-legume mixture and enables it to protect the soil earlier and more

FIG. 178.—Sod on road cut in northeastern Missouri. Sodding in this way is helpful. Note the erosion in the foreground. A slope like this of 25 per cent needs complete protection by sodding all of it. Photographed, near Palmyra, 1936.

completely than if no fertilizer or manure is used. The use of legumes in such a mixture is of utmost importance, because fertilization is not likely to be continued after the vegetation is once established (Fig. 178).

The following mixtures for New York and areas with similar climatic conditions were worked out at the request of the New York Department of Public Works in the fall of 1934 by the writer in cooperation with several fellow workers:

[1] A 5-10-5 fertilizer is a mixture that contains 5 per cent each of nitrogen and water-soluble potash and 10 per cent of available phosphoric acid, or 5, 10, and 5 pounds, respectively, of each constituent in 100 pounds of fertilizer.

Mixture I. For steep grades or sides of cuts which have been surfaced with top soil:

Seed Mixtures	Pounds
Perennial rye grass (*Lolium perenne*)*	20
Timothy (*Phleum pratense*)	10
Kentucky bluegrass (*Poa pratensis*)	25
Rough-stalked meadow grass (*Poa trivialis*)	5
Canada bluegrass (*Poa compressa*)	10
Rhode Island bent grass (*Agrostis tenuis*)	10
Yellow trefoil (*Medicago lupulina*)	5
White Dutch clover (*Trifolium repens*)	5
Redtop (*Agrostis alba*)	10
	100

Mixture II. For road shoulder or any comparatively level places:

Perennial rye grass*	25
Kentucky bluegrass	50
Rhode Island bent grass	20
Kent wild white clover (English Grade A) (*Trifolium repens* var.)	5
	100

Mixture III. For very rough stony or gravelly places or where the application of much good surface soil is expensive, the following mixture may be used:

Hardy hairy or winter vetch (*Vicia villosa*)	10
Mammoth clover (*Trifolium pratense* var.)	10
Sweet clover (*Melilotus alba*)	10
Orchard grass (*Dachylis glomerata*)	20
	50

* Use a winter-hardy variety, preferably *Svalof Victoria.*

From 50 to 100 pounds to the acre of mixtures I and II or 50 pounds of mixture III should give a full stand.

For very dry banks 20 pounds of Canada bluegrass (*Poa compressa*) and 5 pounds of bird's-foot trefoil (*Lotus corniculatus*) to the acre is recommended.

The legume seed should be inoculated immediately before mixing it with the grass seed and planted and covered at once.

The seed should be covered lightly with a hand rake on steep banks or with a weeder on shoulders or wherever a horse can be used. On very steep slopes a light mulch of straw protects the seed and surface soil against being washed away before the sod becomes established. On banks that are not sloped so much as suggested, the use of cheesecloth or burlap serves as protection until the plants become established. The plants come up through the cover which need not be removed.

Adapted grasses and legumes should always be used because they succeed in protecting the soil more quickly and more completely than do unadapted ones. Such legumes are the trailing wild bean (*Strophostyles helvola*) and the partridge pea (*Cassia chamaecrista*). Other locally growing wild ones may be highly

FIG. 179.—Road-cut planting in Virginia. This planting included the trumpet creeper (*Tecoma radicans*), dewberry (*Rubus villosus*), myrtle or common periwinkle (*Vinca Minor*), and black locust (*Robinia Pseudo-acacia*).

useful owing to their ability to make a good growth under unfavorable conditions. Trailing plants such as the wild bean produce a dense cover which protects the soil during the growing period. Vines and shrubs may be planted in the grass-legume mixtures. It is only by so interplanting that the vines and shrubs can protect the soil completely within a reasonable length of time after planting.

Vines.—Several woody vines produce excellent protection. Vines are especially desirable for use on the lower part of road cuts near the actual roadway because they do not interfere with vision along the road. Suitable vines for the Northeast are Hall's

FIG. 180.—Road-cut planting in southern Ohio. This plantation of American bittersweet (*Celastrus scandens*) was made in the fall of 1935. Some slipping occurred at the right, possibly owing to concentration of water there by the diversion ditch.

FIG. 181.—Japanese honeysuckle near Culpepper, *Va*. This old growth of honeysuckle (*Lonicera japonica Halliana*) covers the bank almost completely. It prevents erosion very successfully and its use may well be widely extended.

Japanese honeysuckle (*Lonicera japonica Halliana*), Virginia creeper (*Ampelopsis quinquefolia*), and virgin's bower (*Clematis virginiana*). Some of these plants grow over a very wide range of latitude, particularly the Japanese honeysuckle, which grows well far into the South. The trumpet creeper (*Tecoma radicans*), which grows over a very wide territory including most of the

FIG. 182.—Kudzu controls erosion in highway ditch, in South Carolina. Note how completely the kudzu vines cover the soil. A wide extension of the planting of kudzu in ditches and gullies in the South is highly desirable. (*Courtesy of U. S. Soil Conservation Service, South Carolina.*)

United States east of Texas and Nebraska also, is a useful vine. It is not very common in the northern part of the area north of central Illinois or central New York, an area in which the tips die back in winter (Figs. 179, 180, and 181).

Kudzu may be used on high banks in its habitat if it is kept under control. It is capable of strong growth under unfavorable conditions. For this reason, control of the kudzu is needed in the situations which are favorable for it (Fig. 182).

Fig. 183.—Road-cut planting near Mill Hall, Pa. Several of the shrubs in this trial planting have made good growth. Coralberry (*Symphoricarpos orbiculatus*) has made excellent growth and is holding the slope. Spirea (*billardi*) has survived and made good growth, but some shrubs have suffered considerable winterkilling.

Fig. 184*a*.—Shrubs on left side of road in central New York. Figure 184*a* is directly opposite Fig. 184*b*. The slope varies from 40 to 60 per cent. Virginia creeper is planted nearest the roadway. It is very fitting to plant shrubs on this side since they blend with the adjacent trees.

Shrubs.—In using shrubs on road embankments or cuts, the smaller, lower growing ones are more desirable because plantings must not be permitted to become so tall as to obstruct the view down the highway. Keeping the vision unobstructed is absolutely essential on all trunk and secondary highways but is less important on unimproved dirt roads. Among the low-growing shrubs which may be used in the Northeast are the low bush

FIG. 184*a*.—Grass on right side of road. The slope of this cut is from 60 to 80 per cent. Where the slips occurred the slope is from 100 to 112 per cent. The sod used is Kentucky bluegrass (*Poa pratensis*). Black medic (*Medicago lupulina*) is coming in. The sod was supported by 1- by 2-inch stakes. It appears that a 2-to-1 slope, 50 per cent, would be suitable on this soil. It is peculiarly fitting that grass be used adjacent to pasture.

blueberry (*Vaccinium pennsylvanicum*) and blue myrtle or common periwinkle (*Vinca minor*). Good, slightly taller growing shrubs are the dwarf bush honeysuckle (*Diervilla lonicera*), the coralberry and the buck berry (*Symphoricarpos orbiculatus* and *S. vulgaris*), and the ground hemlock, or Canada yew (*Taxus canadensis*). Somewhat higher growing shrubs which must be confined to the upper part of cuts or the lower part of fills, because they obstruct the vision, are gray stem dogwood (*Cornus paniculata*), red chokeberry (*Aronia arbutifolia*), fragrant sumac (*Rhus canadensis*), high-bush blueberry (*Vaccinium corymbosum*), and pinxter flower (*Azalea nudiflora*).

Larger shrubs may be used if they are more readily available and if they are definitely kept down so as not to interfere with vision along the road. Among these is the hazelnut (*Corylus americana*) which produces a useful nut, silky dogwood (*Cornus amomum*), and red-osier dogwood (*C. stolonifera*), which grow well but must be held down if planted near the roadway (Figs. 183, 184*a* and *b*, 185, 186, and 187).

FIG. 185.—Recently sodded road cut. This road cut in western New York was covered with turf. Slope between 50 and 60 per cent. Such treatment affords excellent protection on well drained soils.

Trees.—Trees are of limited use except on the higher part of deep cuts unless they are pruned regularly. Among them are the black locust (*Robinia Pseudo-acacia*) and hemlock (*Tsuga canadensis*). The merits of the locust have been sufficiently stated, perhaps, but its ability to grow strongly, when inoculated, under conditions unfavorable for most trees and to supply nitrogen for nonlegumes including grasses cannot be overstated. Both the locust and the hemlock stand severe pruning. The black locust sprouts freely from the stump, if cut off; or from the roots, if burned off. It has been used to a considerable extent on road cuts in Pennsylvania.

Many other shrubs and trees can be used to advantage. Adaptation to the locality in which they are planted is highly important from the standpoint of making the desired growth. Wild plants harmonize with their surroundings better than do some of the exotic plants which have been planted for roadside beautification. An important point in favor of local materials is the relatively low cost and the high survival of plants, which are moved only short distances and which are adapted to their surroundings.

Kraebel has described the use of live-wood stakes, cuttings, and wattles in the control of erosion on mountain roads in California.[1]

FIG. 186.—Newly seeded well-sloped cut. This cut has been sloped to approximately 33 per cent. The soil is dry. When the grass becomes established, erosion on this cut will be under control.

The stakes, which are driven into the steep slopes of embankments, fills, or cuts, vary in length from 2 to 4 feet and are set in rows about 3 feet apart across the slope on the contour.

Above the stakes a furrow is made which is filled with bundles of brush of live sprouting wood during the dormant period. For this purpose, willow may be used generally over the United States, and baccharis, in addition, in California. At other times, coarse hay or straw may be used in the furrows (Figs. 188 and 189).

The furrows are nearly but not quite filled with soil which is compacted over the wattle brush in working over it. If too little soil is used, the wattles are undermined and washed out; and if too much, water coming down the slope runs over them. In either case, the wattles fail to serve their purpose.

[1] KRAEBEL, CHARLES J., Erosion Control on Mountain Roads, *U. S. Dept. Agr., Circ.* 380, p. 15, 1936.

Willow wattles sprout and soon hold the soil. During the dormant season live willow is used as stakes. These grow and develop a large root system which holds the soil against slipping and washing. The willow top growth and the grain, grass, and

FIG. 187.—Nature protects a road cut in central New York. Nature needs assistance, however, in order to eliminate erosion quickly. The following plants were found in this natural roadside protection:

Trees:
- White ash (*Fraxinus americana*)
- Black willow (*Salix nigra*)
- Prairie willow (*Salix humilis*)
- American elm (*Ulmus americana*)
- White or silver poplar (*Populus alba*)

Shrubs:
- Buckthorn (*Rhamnus cathartica*)
- Red osier dogwood (*Cornus stolonifera*)
- Black raspberry (*Rubus occidentalis*)
- Choke-cherry (*Prunus virginiana*)
- Staghorn sumac (*Rhus typhina*)

Vines:
- Riverbank grape (*Vitis vulpina*)

Other native and introduced plants:
- Wild carrot (*Dancus carota*)
- Catnip (*Nepeta cataria*)
- Avens (*Geum canadense*)
- Equisetum—horsetail
- New England aster (*Aster novaeangliae*)
- Blue wood aster (*Aster cordifolius*)
- Alfalfa (*Medicago sativa*)
- Timothy (*Phleum pratense*)
- Sweet clover (*Melilotus alba*)
- Canada bluegrass (*Poa compressa*)
- Common burdock (*Arctium minus*)
- Wild strawberry (*Fragaria* spp.)
- Coltsfoot (*Tussilago farfara*)
- Golden rod (*Solidago* spp.)

legumes which are seeded protect the slope against surface washing and gullying. Even without the wattles, stakes of willow and other sprouting woods may be used to good advantage with the grasses, vines, and shrubs previously suggested for the stabilization of road slopes and gully banks.

Kraebel gives an exceptional list of trees, shrubs, vines, grasses, and other plants; methods of management for their production; their place on embankments and cuts; and the elevation and the

FIG. 188.—Willow stakes sprout, in California. This stake produced 23 vigorous sprouts. Willow stakes driven into the soil while dormant sprout and take root. They are highly useful on fills, cuts, and on eroding stream banks. Willows grow over much of the area of the United States. (*Courtesy of C. J. Kraebel, California For. Exp. Sta., U. S. Forest Service.*)

FIG. 189.—Wattle and stake growth on a steep slope in California. This growth was made in 6 months from stakes of common red willow (*Salix laevigata*) set to hold wattles in place. (*Courtesy of C. J. Kraebel, California For. Exp. Sta. U. S. Forest Service.*)

part of California in which each thrives. De France, as indicated at the beginning of this chapter, has worked out a partial list of plants for these purposes in New York. Similar detailed studies are needed for all sections of the country.

Special Reference

Roadside Improvement, by Wilbur H. Simonson and R. E. Royal. *U. S. Dept. Agr., Misc. Pub.* 191, 1936.

CHAPTER XVII

THE CONTROL OF FLOODS

In sections of the East, floods were of unusual severity in 1935 and 1936. In 1935, these floods resulted from rains of such duration and intensity as had not previously been recorded during

FIG. 190.—Flood damage to a state highway bridge. A few hours before the prolonged rains of July 7 and 8, 1935, this was a good concrete highway bridge in Delaware County, New York.

weather bureau history. As shown in Fig. 34, the expected rainfall in central-southern New York in 24 hours once in 100 years is about 5 inches, whereas, according to Fig. 36, the rainfall at points in that area was nearly double that quantity on July 7 and 8, 1935. In March, 1936, the rains were heavy and prolonged, and the soil in open snow-covered fields was frozen. The soil took up little water, and the thawing of the snow by the rain materially increased the total quantity of water pouring down into the valleys. The damage done by such floods as these and the many serious floods in the Midwest and in the West in previous years is incalculable (Figs. 190, 191, and 192).

FIG. 191.—Flood destroys a railroad in New York.

FIG. 192.—Productive soil covered with stones. This field of excellent silage corn was covered with stones to a depth of several feet. Its productivity has been permanently reduced; in fact, the field was all but ruined. In Delaware County, New York.

There are various approaches to, or different views concerning, the ways and means of controlling floods.

Use of Structures.—One viewpoint is that the disastrous effects of floods may be minimized by means of dams for holding runoff water in reservoirs of various types. Dikes or levees along streams have long played a part in flood control.

Dams.—For flood protection above and in the city of Pittsburgh, eleven dams have been proposed.[1] The land and flowage rights required for these projects are reported to be 61,537 acres. This phase of flood control is within the provinces of the hydrographer and the construction engineer; however, there are general considerations which may properly be touched on here.

The function of dams is to impound the excess water and to keep it out of the lower valleys temporarily until that which fell in or near them has run off and is out of the way. Consider a watershed of 30,000 square miles with a rainfall intensity frequency of about 5 inches in 24 hours once in one hundred years. Making provision for 3 inches of runoff, which is not the maximum to be expected, is equivalent to a depth of 1 foot of water over 7,500 square miles. If dams hold 25 feet of water as an average over the area of the reservoir, 300 square miles, or 192,000 acres, of reservoir space is required. If reservoirs keep even 1 inch of rain from the entire watershed out of the lower valleys, marked benefits will have been provided. Holding back the equivalent of 1 inch of water over the entire area of this hypothetical watershed requires 64,000 acres of reservoir area with an average depth of 25 feet, or correspondingly less area under a greater depth of water (Fig. 193).

Most of the land that may be flooded is relatively rich productive soil in the valleys. This area is equivalent to approximately 640 farms having an average of 100 acres each of productive stream-bottom land. And if flood control is developed on this basis generally over the country, this number will be multiplied many times.

Some deposition is bound to take place wherever the velocity of silt-burdened streams is reduced appreciably as it must be even by detention dams. Such deposition, however, is normally much slower than that in the ordinary reservoir. Silting is most

[1] ANGLY, EDWARD, Damming Tomorrow's Floods, *New York Herald Tribune*, Apr. 12, II-1, p. 16. 1936.

rapid in hilly regions that have a relatively large proportion of their areas in intertilled crops and whose growing-season rainfall comes as heavy thundershowers (Fig. 194).

Dikes and Levees.—The building of levees or dikes along the Mississippi River for the protection of farm lands and city property during the past fifty years has been successful only in part. A glance at the soil map of counties bordering on the Mississippi River reveals a large number of breaks in the levees at various times. Breaks in the levee are accompanied by the

Fig. 193.—Flood damage to a suburban home. This lawn is covered with stones to a depth of several feet. The roof of an automobile has been uncovered at the right. The flood water had covered it completely.

deposition of coarse sand. Up to 1905 two breaks in the levee along the Mississippi River bordering Pike County, Illinois, had occurred in a distance of approximately 20 miles.[1] As a result of these breaks, 192 acres of productive land was covered with coarse river sand. In addition, a considerably larger acreage was changed to a mixed sandy loam, much of which is coarse sand of varying depth. These situations came under the writer's observation during 1905 while mapping the soils of Pike County.

The first levees if well built were successful. These levees were not very high, and the pressure on them was not great. As additional drainage districts were formed, and additional levees

[1] Hopkins, C. G., J. G. Mosier, E. Van Alstine, and F. W. Garrett, *Illinois Agr. Exp. Sta., Soil Rept.* 11, 1915.

built, the space for the spreading out of the flood waters was restricted more and more. Higher and higher levees were built as the earlier ones were overtopped and cut out. Moreover, the stream channel tends to silt up during the lower stages of water. The capacity of the channel may thus be so reduced that a subsequent flood overtops the dikes with results that are exceedingly distressing and all too familiar (Fig. 195).

FIG. 194.—A desirable type of small dam. This dam was of the soil-saving type. One upright tile is standing, another fell over when the dam was washed out. Small dams of this nature could impound much water temporarily. Owing to some unusual condition this dam failed.

Most rivers need much of the area of their flood plains on which the flood water may spread out. Consider this situation: a flood plain is 5 miles in width from bluff to bluff. The levees are built rather near the river and allow it an effective width of but 1 mile. At a time when flood water rises to the equivalent of an average depth of 10 feet over the entire flood plain, water approximately 50 feet deep must be carried between the levees. Few levees of any such capacity have been built on the middle section of the Mississippi River. The result of such high water is a break at some weak point in the levee. As the water pours through the break and spreads out over the land formerly protected, the crest of the flood is lowered, and the pressure reduced for the time being. If the water continues to rise, other weak spots break in succession.

Trees and heavy brush on the river side of levees should be held down by cutting. Anything that stands up against the current of the river checks the flow of water and reduces the capacity of the channel including all of the area between the dikes. All sorts of sprouts and fine brush should be encouraged. Such material is overridden by the water, and in fact the brush reduces, if it does not prevent, erosion of the soil between the river bank and the dikes.

Forestry and Agriculture. *Effects of Forests on Flood Control.*—Much evidence which has accumulated shows that a thrifty forest growth with its litter-covered forest floor holds rain water against runoff to a very important extent. This has been shown by means of data from a number of experiment stations in the section on the effects of forests on runoff.

Gibbs found that the soil in unforested areas was frozen to a depth of 20 inches but that in the forest where the snow lay on the litter-covered ground little or no freezing had occurred.[1] At the Northwest Appalachian Erosion Experiment Station, forested and adjacent unprotected areas were compared both during cold winter weather and after the spring thaw. The soil under the forest litter and snow was not frozen so much as in the unprotected soil. Similar conditions were noted during the winter of 1934–1935. Cold weather persisted in January and February, 1936, at Zanesville, Ohio, for a period of two weeks with temperatures as low as −22°. Where the snow was gone or the litter was unusually thin, the soil froze to a depth of 4 to 6 inches. Unpublished data from the U. S. Soil Erosion Experiment Station near Ithaca, N. Y., bear out the Zanesville results on freezing. Gibbs found no runoff from the forested area. Thawing was slight in the forest, but, he says, the water literally poured off the slopes in open fields during the flood period of March, 1936.

Attention is called to Fig. 42. This picture was taken about 15 hours after the torrential rain of July 7 and 8, 1935, had ceased. No water was running from cultivated fields or pastures at that time. The runoff from them was already in the streams crowding along on its way to the sea. The forest on the top of the ridge above had delayed the runoff from the heavy downpour until after the water from other lands was well on its way. In other

[1] Gibbs, J. A., Forest Litter, Freezing and Run-off, *U. S. Dept. Agr., Soil Conservation*, Vol. I, No. 11, p. 14, 1936.

words, the crest of the flood and the flood damage was held down by the forest. It is characteristic of forests to absorb ordinary heavy rains with no runoff whatever.

Not only do forests reduce the height of floods, but they hold back the water and cause it to soak into the soil from which it reappears later as springs, thereby evening up stream flow by increasing it between rains. Thus, reforestation of hilly lands unsuited to cultivation not only aids in flood control but aids in supplying water during dry periods for navigation and for the production of water power.

FIG. 195.—Levee protecting farm land along the Illinois River. Note the willows on the river side of the levee. Keeping these down permits the water to get away and reduces the crest of floods locally. But while doing this it hastens the attainment of and raises the crest of floods on the Mississippi.

Responsibility for Erosion on Publicly Owned Lands.—It appears to be the duty of any public agency—nation, state, county, city, town, or school district—that takes title to land to assume responsibility for and eliminate soil erosion on such lands at the earliest possible moment. Full cooperation between all divisions of government and the owners of private property is highly desirable for the ultimate elimination of flood losses.

Land taken out of cultivation does not rapidly reclothe itself with vegetation so as to control the loss of soil and water from the land. Control may be brought about by seeding grasses and legumes with or without a nurse crop. In some sections of the country light fertilization on badly eroded or worn lands is essential in order to assist the seeding in reducing or eliminating soil and water losses immediately. Trees planted in such seedings are benefited rather than injured, and erosion is controlled during their early growth.

Contour furrows if carefully laid out and well made assist greatly in holding water until it soaks into the soil, and consequently such furrows reduce soil erosion and alleviate flood conditions.

Effects of Agriculture on Flood Control.—One point for early attack on flood-control problems is at the tops and on the sides of ridges at the source, that is, on the farms or on idle land. Effort directed at holding every raindrop on the land where it falls is relatively effective. *Good soil-erosion control is fundamental in successful flood control.* Steep slopes are seeded to pasture or given pasture treatment. The result is almost complete control of erosion and absorption of a very high percentage of the rainfall. Somewhat less steep land may be seeded to permanent meadow or to alfalfa on soils adapted to that crop. Like pasture, these close-growing crops essentially prevent erosion and hold much of the water that falls on them except during the very heaviest downpours. And even then, much of the water is taken up by the soil, or at least its flow is retarded to such an extent that it does not reach the main stream until the peak of floods has passed.

On less steep slopes the growing of small grain and intertilled crops in a long rotation on the contour in strips goes a long way toward reducing erosion and loss of water from the land. Diversion ditches are used to take water off slopes and spread it out on forest or grass land where much of it is taken up by the soil. Good soil-erosion control farming conserves water and at the same time conserves the soil. And as it is effective in controlling erosion and loss of water, so it is effective as an aid in flood control.

The terraces used on farm lands in the South, the Southwest, the Midwest, and the Northwest are forms of diversion, or hill-side ditches the purpose of which is to slow down the runoff of rain water and to encourage its absorption by the soil. In the dry areas every drop of water is needed in the soil to produce a vegetative cover for the control of soil movement by blowing and for the production of crops. Unfortunately, many dry-land areas receive a high proportion of the year's precipitation as torrential rains. This leads to heavy losses of water and soil. Level terraces and contour furrows are used in meadows and pastures to hold the water and to cause its absorption by the soil. Water absorbed does not reach the streams until long after the flood stage has passed.

Every drop or every cubic foot of water held back on the farms or in the forests by good farming and by good forest practice is one unit of water deducted from the crest of floods.

Conclusion.—Dams, unquestionably, are helpful in controlling floods during the period of their full or major effectiveness. Much control of soil erosion will be required to slow down the rate of silting of such reservoirs. Some risk to the populations and the property below the dams may be involved in their construction. Flood control appears from the foregoing to be dependent in a greater measure than is commonly considered on the reforestation of steep lands and on soil-erosion control on good sloping farm lands—briefly, on good farming and on good forest practice.

Controlling erosion brings about absorption by the soil of much water which otherwise is lost as runoff and which raises the crest of floods. Such absorbed water reappears later as springs and feeds the streams with a marked degree of regularity. Thus soil conservation leads to water conservation. Water conservation tends toward uniformity of stream flow and, therefore, improves conditions for fish, for navigation, and for the production of water power. In time, the control of soil erosion may be expected to reduce materially the damage now resulting from floods.

INDEX